Data Analysis for Chemistry

DATA ANALYSIS FOR CHEMISTRY

An Introductory Guide for Students and Laboratory Scientists

D. Brynn Hibbert

J. Justin Gooding

UNIVERSITY PRESS
2006

OXFORD
UNIVERSITY PRESS

Oxford University Press, Inc., publishes works that further
Oxford University's objective of excellence
in research, scholarship, and education.

Oxford New York
Auckland Cape Town Dar es Salaam Hong Kong Karachi
Kuala Lumpur Madrid Melbourne Mexico City Nairobi
New Delhi Shanghai Taipei Toronto

With offices in
Argentina Austria Brazil Chile Czech Republic France Greece
Guatemala Hungary Italy Japan Poland Portugal Singapore
South Korea Switzerland Thailand Turkey Ukraine Vietnam

Published by Oxford University Press, Inc.
198 Madison Avenue, New York, New York 10016
www.oup.com

Library of Congress Cataloging-in-Publication Data
Hibbert, D. B. (D. Brynn), 1951–
Data analysis for chemistry: an introductory guide for students and laboratory scientists/
D. Brynn Hibbert and J. Justin Gooding.
p. cm.
ISBN-13: 978-0-19-516210-3; 978-0-19-516211-0 (pbk.);
0-19-516210-2; 0-19-516211-0 (pbk.)
1. Chemistry–Statistical Methods. 2. Analysis of variance. I. Gooding, J. Justin
II. Title.
QD39.3.S7H53 2005
540′.72–dc22 2004031124

9 8 7 6 5 4 3 2 1
Printed in the United States of America
on acid-free paper

This book is dedicated to the legion of students that have passed through Schools of Chemistry who have tried to unravel the mysteries of data analysis.

Preface

The motivation for writing this book came from a number of sources. Clearly, one was the undergraduate students to whom we teach analytical chemistry, and who continually struggle with data analysis. Like scientists across the globe we stress to our students the importance of including uncertainties with any measurement result, but for at least one of us (JJG) we stressed this point without clearly articulating how. Conversations with many other teachers of science suggested JJG was not the exception but more likely the rule. The majority of lecturers understood the importance of data analysis but not always how best to teach it. In our school, like many others it seems, the local measurement guru has a good grasp of the subject, but the rest who teach other aspects of chemistry, and really only use data analysis as a tool in the laboratory class, understand it poorly in comparison. This is something we felt needed to be rectified, a second motivation.

In conversation between the pair of us we came to the conclusion that the problem was partly one of language. In writing this book we also came to the conclusion that another aspect of the problem was the uncertainty that arises from any discipline which is still evolving. Chemical data analysis, with aspects of metrology in chemistry and chemometrics, is certainly an evolving discipline where new and better ways of doing things are being developed. So this book tries to make data analysis simple, a sort of idiot's guide, by (1) demystifying the language and (2) wherever possible giving unambiguous ways of doing things (recipes). To do this we took one expert (DBH) and one idiot (JJG) and whenever DBH stated what should be done JJG badgered him with questions such as, "What do you mean by that?," "How exactly does one do that?," "Can't you be more definite?," "What is a rule of thumb we can give the reader?" The end result is the compromise between one who wants essentially recipes on how to perform different aspects of data analysis and one who feels the need to give, at the very least, some basic information on the background principles behind the recipes to be performed. In the end we both agree that for data analysis to be performed properly, like any science, it cannot be treated as a black box but for the novice to understand how to perform a specific test how to perform it must be unambiguous.

So who should use this book? Anybody who thinks they don't really understand data analysis and how to apply it in chemistry. If you really do understand data analysis, then you may find the explanations in the book too simple and the scope too limited. We see this as very much an entry level book which is targeted at learning and teaching undergraduate data analysis. We have tried to make it easy for the reader to find the information they are seeking to perform the data analysis they think they need. To do this we have put the glossary at the beginning of the book with directions to where in the book a certain concept is located. We also add in this initial Readers' Guide frequently asked questions (FAQs) with brief answers and directions to where more detailed answers are located, and a list of useful Microsoft Excel functions. Hopefully together these three sections will help you find out how to do things like when your lecturer tells you to "measure a calibration curve and then determine the uncertainty in your measurement of your unknown." If after looking through this book, and then sitting down to work through the examples, you still are saying "How?" then we haven't quite achieved our objective.

Acknowledgments

First and foremost we would like to thank our families for the neglect they suffered as we wrote this book. In particular Marian, Hannah, and Edward for DBH and Katharina for JJG.

We would also like to thank the members of our research group for the neglect they also suffered as a result of us being diverted by this project. Some of them repaid us for that neglect by carefully reading through the manuscript and making many suggestions so a very big thank you goes to Dr. Till Böcking, Dr. Florian Bender, and soon to be Doctors Edith Chow and Elicia Wong.

We would also like to thank our colleagues in the School of Chemistry at the University of New South Wales and beyond for help.

Finally we would like to thank the students to whom this book is dedicated for their questions and their hard work in trying to understand this sometimes baffling subject.

Spreadsheets and screenshots are reproduced with permission from Microsoft Corporation.

Contents

Data Analysis for Chemistry

Readers' Guide: Definitions, Questions, and Useful Functions

Where to Find Things and What to Do

This chapter is called Readers' Guide because chapter 1 is clearly the proper start of the book, with introductions and discussions of what measurement really is and so on. This chapter was compiled last, and attempts to be the first stop for a reader who does not want the edifying discourse on measurement, but is desperate to find out how to do a t-test. In the glossary, we define terms and concepts used in the book with a section reference to where the particular term or concept is explained in detail. If you half know what you are after, perhaps the memory jog from seeing the definition may suffice, but sometime return to the text and reacquaint yourself with the theory.

There follows "frequently asked questions" that represent just that—questions we are often asked by our students (and colleagues). The order roughly follows that of the book, but you may have to do some scanning before the particular question that is yours springs out of the page.

Finally we have lodged a number of Excel spreadsheet functions that are most useful to a chemist faced with data to subdue. The list has brought together those functions that are not obviously dealt with elsewhere, and does not claim to be complete. But have a look there if you cannot find a function elsewhere.

Glossary

The definitions given below are not always the official statistical or metrological definition. They are given in the context of chemical analysis, and are the authors' best attempt at understandable descriptions of the terms.

α The fraction of a distribution outside a chosen value. (Section 2.5.2)

Accuracy Formerly: the closeness of a measurement result to the true value; now: the quality of the result in terms of trueness and precision in relation to the requirements of its use. (Section 1.8; figure 1.6)

Analytical sensitivity The linear coefficient representing the slope of the relationship between the instrument response and the concentration of standards. In other words, the slope of the calibration plot. (Section 5.3)

ANOVA (analysis of variance) A statistical method for comparing means of data under the influence of one or more factors. The variance of the data may be apportioned among the different factors. (Chapter 4)

Arithmetic mean $\bar{x}$ The average of the data. The result of summing the data and dividing by the number of data (n). (Section 2.4.1)

Bias A systematic error in a measurement system. (Section 1.7)

Calibration The process of establishing the relation between the response of an instrument and the value of the measurand. (Section 5.2)

Calibration curve A graph of the calibration. (Section 5.2)

Central limit theorem The distributions of the *means* of n data will approach the normal distribution as n increases, whatever the initial distributions of the data. (Section 2.4.6)

Certified reference material (CRM) A standard with a quantity value established to a high metrological degree, accompanied by a certificate detailing the establishment of the value and its traceability. Used for calibration to ensure traceability, and for estimating systematic effects. (Section 3.3)

Confidence interval A range of values about a sample mean which is believed to contain the population mean with a stated probability, such as 95% or 99%. The 95% confidence interval about the mean ($\bar{x}$) of n samples with standard deviation s is: $\bar{x} \pm t_{0.05'',n-1}(s/\sqrt{n})$. $t_{0.05'',n-1}$

is the 95%, two-tailed Student t-value for $n-1$ degrees of freedom. (Section 2.5.1)

Confidence limit The extreme values defining a confidence interval. (Section 2.5.1)

Correction for the mean Subtraction of the *grand mean* from each measurement result in ANOVA. This quantity is also known as the *mean corrected value*. (Section 4.4)

Corrected sum of squares See total sum of squares. (Section 4.4)

Cross-classified system In a multiway ANOVA when the measurements are made at every combination of each factor. (Section 4.8)

Degrees of freedom The number of data minus the number of parameters calculated from them. The degrees of freedom for a sample standard deviation of n data is $n-1$. For a calibration in which an intercept and slope are calculated, $df = n-2$. (Sections 2.4.5, 5.3.1)

Dependent variable The instrument response which depends on the value of the independent variable (the concentration of the analyte). (Section 5.2)

Detection limit See limit of detection. (Section 5.8)

Effect of a factor How much the measurand changes as a factor is varied. (Section 4.3)

Error The result of a measurement minus the true value of the measurand. (Section 1.7)

Factor In ANOVA a quantity that is being investigated. (Sections 4.2; 4.3)

Fisher *F*-test A statistical significance test which decides whether there is a significant difference between two variances (and therefore two sample standard deviations). This test is used in ANOVA. For two standard deviations s_1 and s_2, $F = s_1^2/s_2^2$ where $s_1 > s_2$. (Sections 3.7, 4.4)

Fit for purpose The principle that recognizes that a measurement result should have sufficient accuracy and precision for the user of the result to make appropriate decisions. (Section 1.10)

Grand mean The mean of all the data (used in ANOVA). (Section 4.2)

Gross error A result that is so removed from the true value that it cannot be accounted for in terms of measurement uncertainty and known systematic errors. In other words, a blunder. (Section 1.7)

Grubbs's test A statistical test to determine whether a datum is an outlier. The G value for a suspected outlier can be calculated using $G = (|x_{suspect} - \bar{x}|/s)$. If G is greater than the critical G value for a stated probability ($G_{0.05'',n}$) the null hypothesis, that the datum is not an outlier and belongs to the same population as the other data, is rejected at that probability. (Section 3.5)

Heteroscedastic data The variance of data in a calibration is not independent of their magnitude. Usually this is seen as an increase in variance with increasing concentration (e.g., when the relative standard deviation is constant for a calibration). (Section 5.3.1)

Homoscedastic data The variance of data in a calibration is independent of their magnitude (i.e., the standard deviation is constant). (Section 5.3.1)

Hypothesis test Where a question about data is decided upon based on the probability of the data given a stated hypothesis. (Section 3.1)

Independent measurements Measurements made on a number of individually prepared samples. (Section 2.7)

Independent variable A quantity that is under the control of the analyst. In calibration, it is the quantity varied to ascertain the relationship between this quantity and the instrumental response. Typically in a calibration model the independent variable is concentration. (Section 5.2)

Indication of a measuring instrument The instrumental response or output. (Section 5.3)

Indication of the blank The instrumental response to a test solution containing everything except the analyte. If this is not possible to measure, it may taken as the intercept of the calibration curve. (Section 5.3)

Influence factor (quantity) Something that may affect a measurement result. For example, temperature, pressure, solvent, analyst. In calibration, influence quantities refer to quantities that are not the independent variable but that may affect the measurement. (Sections 4.2, 4.3, 5.3)

Instance of factor Particular example of a factor in an ANOVA. For example, in an experiment performed at 20, 30, and 40°C, the three temperatures are instances of the factor "temperature." (Section 4.2)

Interaction In a multiway ANOVA an effect of one factor on the effect of another factor on the response. For example if a reaction rate is increased more by an increase in temperature at short reaction times than longer reaction times, then there is said to be a "temperature by time" interaction. (Section 4.8)

Intercept The constant term in a calibration model. See indication of blank. (Section 5.3)

Interquartile range The middle 50% of a set of data arranged in ascending order. The *normalized interquartile range* serves as a robust estimator of the standard deviation. (Section 2.6.2)

Intralaboratory standard deviation The standard deviation of measurement results obtained within the same laboratory but not under *repeatability conditions*, for example by different analysts using different equipment on different days. (Section 2.7)

Leverage The tendency of a single point to drag the calibration line towards it and hence increase the value of the standard error of the regression ($s_{y/x}$). (Section 5.3.1)

Limit of detection Smallest concentration of analyte giving a significant response of the instrument that can be distinguished above the blank or background response. (Section 5.8)

Limit of determination The smallest value of a measurand that can be measured with a stated precision. (Section 5.8)

Linear calibration model Equation for the instrumental response which is directly proportional to the concentration (of the form $y = a + bx$). (Section 5.3)

Linear range The region in a calibration curve where the relationship between instrumental response and concentration is sufficiently linear for its use. (Section 5.3.2)

Mean (population mean) μ The average value of the data set which defines the probability density function. The population mean is the true value in the absence of systematic error. (Section 1.8.2)

Mean (sample mean) $\bar{x} = \left(\sum_{i=1}^{1=n} x_i/n\right)$ The *arithmetic mean* of a data set. The result of summing the data and dividing by the number of data (n). (Section 2.4.1)

Mean square A sum of squares divided by the degrees of freedom. (See residual sum of squares, sum of squares due to the factor studied.)

Means *t*-test t-test to decide if two sets of data come from populations having the same mean. For each set calculate the sample mean and standard deviation ($\bar{x}_1, s_1, \bar{x}_2, s_2$). Test the standard deviations under the hypothesis $\sigma_1 = \sigma_2$ (see F-test). If the populations have equal variance, $t = (|\bar{x}_1 - \bar{x}_2|/s_p\sqrt{1/n_1 + 1/n_2})$ where $s_p^2 = ((n_1 - 1)s_1^2 + (n_2 - 1)s_2^2)/(n_1 + n_2 - 2)$ and degrees of freedom $n_1 + n_2 - 2$. If the populations have unequal variance, $t = (|\bar{x}_1 - \bar{x}_2|/\sqrt{S_1^2/n_1 + S_2^2/n_2})$ with degrees of freedom

$$\frac{(s_1^2/n_1 + s_2^2/n_2)^2}{(s_1^4/n_1^2(n_1 - 1)) + (s_2^4/n_2^2(n_2 - 1))}. \quad \text{(Section 3.8)}$$

Measurand The quantity that is intended for measurement. (Section 1.7)

Measurement Set of operations having the object of determining the value of a quantity. (Section 1.2)

Measurement uncertainty A property of a measurement result that describes the dispersion of values that can be attributed to the measurand. It quantifies our confidence in a measurement result. (Section 1.7.3)

Median The middle value of a set of data arranged in order of magnitude. (Section 2.6.1)

Multivariate calibration Calibration in which multiple independent variables are used to establish the calibration model. (Section 5.2)

Nested factor In multiway ANOVA a factor that is varied separately for each level of another factor. (Section 4.8)

Normal (Gaussian) distribution The random distribution described by the probability density function which gives the familiar "bell-shaped curve." It is described by the mean μ and standard deviation σ $f(x|\mu,\sigma) = (1/\sigma\sqrt{2\pi})\exp\left[-((x-\mu)^2/2\sigma^2)\right]$. (Section 1.8.2)

Null hypothesis (H_0) The hypothesis that the population parameters being compared (e.g., mean or variance) on the basis of the data are the same, and the observed differences arise from random variation only. This is the hypothesis used in many statistical significance tests that "there is no difference between the factors that are being compared." (The null hypothesis is first introduced in section 3.2 but is used throughout chapters 3 and 4). (Section 3.2)

One-way ANOVA an ANOVA in which a single factor is varied. (Section 4.4)

Outlier A datum from a sample, assumed to be normally distributed, which lies beyond the mean at a stated probability. Therefore, an outlier is a datum that, according to a statistical test, does not belong to the distribution of the rest of the data. (Section 3.5)

Paired *t*-test A statistical significance test for comparing two sets of data where there are no repeat measurements of a single test material but there are single measurements of a number of different test samples. To perform this test you use $t = (|\bar{x}_d|\sqrt{n}/s_d)$ where $\bar{x}_d$, s_d are the mean and standard deviation of n differences. (Section 3.9)

Population The infinite number of results that could be obtained in an experiment that are described by the probability density function. (Section 2.3)

Precision The standard deviation of measurement results obtained under specified conditions (see repeatability, reproducibility). (Section 1.8; figure 1.6)

Probability density function (pdf) The mathematical function that describes a distribution in terms of the probability of finding a result. For the *normal distribution* the pdf is the "bell-shaped curve." (Section 1.8.2; equation 1.1)

Quantity Attribute or phenomenon, body or substance that may be distinguished qualitatively and determined quantitatively. (Section 1.4)

Q-test (Dixon's Q-test) An outlier test. Grubbs's test is the preferred test to use. (Section 3.5)

Random error Variation in the quantity measured with repeated measurements centered around the true value. It is described by the *normal distribution.* (Section 1.7)

Regression The process of determining the optimum parameters of a model that fit some data. For example, given pairs of data (x, y) a linear model finds the best fit values of the intercept (a) and slope (b) in $y = a + bx$. Least squares regression minimizes the sum of the squares of the *residuals.* (Section 5.3.1)

Relative standard deviation (RSD) The sample standard deviation expressed as a percentage of the mean, RSD $= 100 \times \frac{s}{\bar{x}}$. Also called the coefficient of variation (CV). (Section 2.4.3)

Repeatability The precision of an analytical method, usually expressed as the standard deviation of independent determinations performed by a single analyst on the same day using the same apparatus and method. (Section 2.7)

Reproducibility The precision of an analytical method, usually expressed as the standard deviation of determinations performed in different laboratories (and therefore by different analysts using different equipment on different days). (Section 2.7)

Residual $(y_i - \hat{y}_i)$: the difference between the measured response y_i and the response estimated from the regression equation for the calibration curve $(\hat{y}_i)$. (Section 5.3.1)

Residual sum of squares, SS_r Also called "within variables sum of squares," is the difference between the *total sum of squares* and the *sum of squares due to the factor studied.* This number is used in determining whether there is a significant difference between two means using ANOVA. (Section 4.4)

Robust estimator Estimators of parameters of the distribution of data that can tolerate extreme values (outliers). (Section 2.7)

Sample Statistically this is the set of n data being investigated. (Section 2.3)

Significance test A statistical test to determine whether there is a statistically significant difference between two sets of data at a defined probability level. (Section 3.2)

Slope See analytical sensitivity. (Section 5.3)

Standard addition A method of analysis in which a measurement is made on the sample followed by a second measurement after a known amount of calibration material is added to the sample. (Section 5.7)

Standard deviation (population standard deviation), σ The square root of the variance, the population standard deviation represents the dispersion of the population. In the *normal distribution*, 68% of the distribution lies at the mean $\mu \pm 1\ \sigma$. (Section 1.8.2)

Standard deviation (sample standard deviation), s An estimate of σ from n data calculated as $\sqrt{\left(\sum_{i=1}^{i=n}(x_i - \bar{x})^2\right)/(n-1)}$. (Section 2.4.2)

Standard deviation of the mean (σ_n) The standard deviation of means of n data. It is related to the standard deviation of the population (σ) by $\sigma_n = \sigma/\sqrt{n}$. The sample standard deviation of the mean is estimated from $s/\sqrt{n}$. (Section 2.4.6)

Standard error of the regression ($s_{y/x}$) A quantity that is a measure of the goodness of fit of a regression equation for a calibration curve: $s_{y/x} = \sqrt{\left(\sum_{i=1}^{i=n}(y_i - \hat{y}_i)^2\right)/df}$, where $(y_i - \hat{y}_i)$ is the *residual* of the point i and df are the *degrees of freedom*. The better the fit the smaller $s_{y/x}$. (Section 5.3.1)

Student's *t*-test, Student *t*-value See *t*-test.

Sum of squares due to the factor studied, SS_c Also known as treatment sum of squares, heterogeneity sum of squares, or between column sum of squares. It is a quantity in ANOVA which is related to the variance between factors. (Section 4.4)

Systematic error A deviation from the true value that is always of the same magnitude and in the same direction from the mean. It should be estimated from measurement of *certified reference materials* and corrected for in a chemical analysis. Significant systematic error can be tested using $t = \left(\left|x_{\text{assigned}} - \bar{x}\right|\sqrt{n}/s,\right)$ where n independent measurements of a reference material with assigned value x_{assigned} have been made giving mean $\bar{x}$ and standard deviation s. (Sections 1.7, 3.3, 3.6)

Tails In a normal distribution the bell curve is symmetrical about the mean. The values either side of the mean, that is the parts of the bell curve greater than and less than the mean are the "tails" of the probability distribution function. (Section 2.5.4)

Test material The actual material being studied. For example, if the concentration of a solution is being analyzed it is called a test solution, if it is an extract that is being analyzed it is a test extract. The use of the word sample is not encouraged because of confusion with the statistical concept of a *sample*. (Section 2.3)

Total sum of squares, SS_T (also corrected sum of squares) In ANOVA the number arising from the sum of the squares of the mean corrected values. (Section 4.4)

***t*-test** (Student's *t*-test) A statistical significance test for hypotheses concerning the mean of a small sample. A *t*-value is calculated (t_{calc}) and the probability that this *t*-value would be exceeded in a great number of replicate measurements is obtained, $p(T > t_{\text{calc}})$. The tested hypothesis is then accepted or rejected on the basis of the probability. See also means *t*-test. (Section 3.8)

Type I error (false positive) Rejecting a hypothesis when it is true. In terms of the null hypothesis this means the significance test shows there is a difference in the two sets of data but in fact there is no difference. (Section 3.3)

Type II error (false negative) Accepting a hypothesis when it is false. This means the significance test shows there is no difference between the data being compared but in fact there is. Another way of saying this is the test suggests the null hypothesis is correct but actually it is incorrect. (Section 3.3)

Univariate calibration When only one *independent variable* is being used to establish the relation between the instrument response and the value of the measurand. (Section 5.2)

Value Magnitude of a particular quantity generally expressed as a number multiplied by a unit of measurement. (Section 1.4)

Variance (population), σ^2 the square of the population standard deviation. (Section 1.8.2)

Variance (sample), s^2 The square of the sample standard deviation. (Section 2.4.2)

$\hat{x}$ The estimated concentration of an unknown determined using a calibration. (Section 5.3)

$z_{\alpha/2}$ The number of standard deviations either side of the mean containing a fraction 1 α of the distribution. (Section 2.5.2)

z-score The number of standard deviations a data point is from the mean. It is often used in significance testing such as testing for a suspected outlier. (Section 2.5.2)

Frequently Asked Questions (FAQs)

1. Why should I bother with data analysis anyway?
   Unless you are just going to tabulate all the results you have and not make any conclusions, then you need some way to treat your results to deliver information to whoever is interested in your doing the experiment in the first place. (Chapter 1)
2. Why bother with uncertainties?
   Because an analytical result without information regarding the uncertainty of the value is useless. (Section 1.6)
3. What is the difference between the measurand and the analyte?
   The measurand is the *quantity* that is being measured. For example, the concentration of dioxin in drinking water is the measurand. The analyte is the dioxin (and the matrix is the drinking water). (Section 1.4)
4. What is the difference between precision, standard deviation, and uncertainty?
   Precision is a measure of the variability of results obtained under different circumstances (e.g., repeatability or reproducibility). It is usually expressed as a standard deviation. Uncertainty is a general concept that covers all aspects of our lack of knowledge of the true value. It is assessed by an "uncertainty budget" and also is expressed in terms of a standard deviation. (Sections 1.7, 1.8)
5. How do I make my measurements *traceable* to an international standard such as the SI?
   By calibrating using traceable standards such as certified reference materials. (Section 1.5)

6. Can I use data analysis to tell me why an error has occurred?
   No! It can, however, allow you to identify systematic error and determine the uncertainty as a consequence of random error. (Sections 1.7, 3.6)
7. When writing an uncertainty, how many significant figures should I use?
   If the standard deviation or 95% confidence level is known then write this value to 2 significant figures. The measurement result can then be written to the same number of decimal places. For example: (1.123 ± 0.032) M. (Section 1.9.2)
8. Is my uncertainty reasonable? What uncertainty is acceptable?
   There is no simple answer to this! It all depends on what the answer will be used for and how much time you have. Essentially you must make a measurement with sufficient accuracy to allow appropriate decisions to be made. This is known as "fit for purpose." (Section 1.10)
9. What is "fit for purpose?"
   Making a measurement with sufficient accuracy to allow appropriate decisions to be made. (Section 1.10)
10. Why is it necessary to perform repeat measurements?
    The more repeats that are done, the smaller the uncertainty in the sample mean and hence the more confident one becomes that the sample mean is a good estimator of the population mean. (Section 2.4)
11. When calculating the standard deviation on my calculator which button do I use—the σ (also written as $x\sigma_{\text{n}}$ on some calculators) or the $s(x\sigma_{\text{n}-1})$?
    Always use $s(x\sigma_{\text{n}-1})$ which gives the sample standard deviation. (Example 2.1a)
12. When do I quote a variance and when do I quote a standard deviation?
    As the variance is the square of the standard deviation, either gives equivalent information. However, as the standard deviation has the same units as the measurand, it may be more obviously interpreted. (Section 2.4.2)
13. What is the abbreviation for standard deviation (s, sd, SD, σ)?
    A sample standard deviation is s. The population standard deviation is σ. (Section 2.4.2)
14. What are the units of standard deviation?
    The same as the units of the mean. (Section 2.4.4)

15. Why quote a relative standard deviation (RSD) rather than a standard deviation?
    It gives an immediate impression of the precision of the measurement without knowledge of the value of the quantity. (Section 2.4.3)
16. Data analysis seems to be based on a large number of data points and a normal distribution. What if I only have a few points?
    You can still use the statistical approaches outlined in this book by assuming the data are normally distributed. However, the uncertainty in the estimates of mean and standard deviation are increased and there does come a point that there is little to be gained from calculation of these parameters (say with $n < 6$). (Section 1.8.2)
17. What happens to the standard deviation and the standard deviation of the mean as the number of data increases?
    The sample standard deviation (s) approaches the population standard deviation (σ). The standard deviation of the mean approaches zero. (Sections 2.4.2, 2.4.6)
18. Once I have determined the mean and standard deviation can I quote the results as $\bar{x} \pm s$?
    No, $\pm$ should be reserved for confidence limits with stated coverage (e.g., 95%). If you want to quote the mean with a standard deviation then write as $\bar{x}$ (s = standard deviation, n = number of data). (Section 2.5)
19. After how many measurements can I assume $s = \sigma$?
    After 30 measurements the error in taking s for σ is about 4%. (Section 2.5.3)
20. When should I use robust estimators?
    If you are concerned that the data are not normally distributed or have extreme outliers, robust estimators such as the median and interquartile range may be more useful. (Section 2.6)
21. What do you mean by one tailed and two tailed?
    The normal distribution is symmetrical about the mean. When we talk about a certain percentage of the distribution we can choose the area from infinity which leaves the remaining area at one end (one tailed), or the area either side of the mean, leaving half the remaining area at either end of the distribution (two tailed). (Section 2.5.4)

22. How can I test whether my data are normally distributed?
    If you have enough data you can plot a histogram and decide if it appears suitably bell shaped. A Rankit plot is also a useful visual test of normality and may be used with fewer data. (Sections 1.7.2, 3.4)
23. If my data are not normally distributed how do I estimate a mean and an uncertainty?
    See FAQ 20.
24. When performing a significance *t*-test what probability level do I set the null hypothesis to be rejected?
    It all depends on for what purpose the data will be used. Commonly 95% or 99% are used but you should consider the risk of making a Type I or Type II error. (Section 3.2)
25. How do I determine whether a datum is an outlier?
    Perform a Grubbs's test. (Section 3.5)
26. How many data can I assign as outliers using the Grubbs's test given in chapter 3?
    Only one. There is a Grubbs's test for pairs of outliers (Massart et al.—see Bibliography). Any more and you should be asking yourself whether the data is normally distributed. (Sections 3.4, 3.5)
27. When can I discard data?
    Never. You may decide not to use a value in the calculation of mean and standard deviation after performing a Grubbs's test for an outlier. (Section 3.5)
28. What is a one-tailed significance test, and what is a two-tailed significance test?
    A significance test rejects the null hypothesis when the probability of the test statistic falls below a given value (e.g., $\alpha < 0.05$ for a 95% test). A one-tailed test has all this probability at one end of the distribution only. A two-tailed test has half the probability at one end of the distribution and half at the other. (Section 3.6)
29. So when should I use a one- or two-tailed test?
    When you are testing two means use a two-tailed test when you have no reason to believe one is bigger or smaller than the other. Use a one-tailed test if you want to know if one mean is significantly greater than the other. (Section 3.6)
30. What is α?

α is a probability, between 0 and 1, of a particular test statistic given the hypothesis being tested, at which the hypothesis is rejected. For example, $\alpha = 0.05$ means we reject the hypothesis when the probability of finding the data given the hypothesis falls below 5%—a so-called 95% test. (Sections 2.5.2, 3.6)

31. What is the difference between α', α'', and $\alpha/2$ in a significance test?
    $\alpha/2$ implies a one-tailed test, also written α'. α'' refers to a two-tailed test with $\alpha/2$ at either end of the distribution. For example, $t_{0.05'',\mathrm{df}} = t_{0.025',\mathrm{df}} = t_{(0.05/2),\mathrm{df}}$. (Section 3.6)
32. How can I decide whether one analytical method is better than another?
    "Better" is a question that relates to what the measurement result will be used for. However, an estimate of the precisions of each method and whether there is systematic error are important. (Sections 3.6, 3.7)
33. When do I do a means *t*-test and when do I do a paired *t*-test?
    When there are a number of repeated analyses of the same material then do a means *t*-test. When there are many different test materials with a single measurement performed, then do a paired *t*-test. (Sections 3.8, 3.9)
34. Can I test for bias without a sample of known value?
    No, if you only have your method with which to do the analysis. (Section 3.6)
35. What is the difference between *recovery* and *bias*?
    They are both types of *systematic error*. Bias usually refers to systematic error in an instrument and is an absolute difference. Recovery is the fraction of an analyte that is presented to the measuring instrument. It is often less than 100% because of losses during preparation of the test material before measurement. Both may be estimated and corrected for. (Sections 1.7, 3.6)
36. How do I avoid making Type I errors (reject H_0 when it is true)?
    Decrease α. That is, test at greater probability levels (95, 99, 99.9%. etc.). (Section 3.3)
37. How do I avoid making Type II errors (accept H_0 when it is false)?

Increase α. That is, test at lesser probability levels (95, 90, 80%, etc.). (Section 3.3)

38. So how do I choose what probability to use?
Think about the relative risk of making Type I and Type II errors. (Section 3.3)
39. How do I know whether two analytical methods give equivalent results or not?
Test the means of results of analyses by each method of aliquots of a test material by a *t*-test. (Section 3.8)
40. When should I use ANOVA and when a *t*-test?
Use ANOVA if you want to know if there is significant difference among a number of instances of a factor. Always use ANOVA for more than one factor. ANOVA data must be normally distributed and homoscedastic. Use a *t*-test for testing pairs of instances. The data must be normally distributed but need not be homoscedastic. (Sections 3.8, 4.2)
41. When optimizing an analytical method how do I determine which variables cause a significant change to the method performance?
Do an ANOVA which allows you to look at the variance in the data, use the *p*-value from the ANOVA results table decide if there is a significant effect caused by a factor. (Section 4.2)
42. Can I do ANOVA with different numbers of replicates of an instance of a factor in Excel?
Yes, for single-factor ANOVA. No, for two-factor ANOVA. (Sections 4.6, 4.9)
43. What is a factor and what is an instance of a factor?
A factor is whatever we are testing in ANOVA, for example an analytical method, sampling position in a silo, the gender of an analyst. Instances of the factor are the particular examples of that factor chosen for study, for example a spectrophotometric method and an electrochemical method, measures at the top, middle, and bottom of a silo, and male analysts and female analysts. (Section 4.3)
44. If I find a significant difference between factors how can I determine which factor or factors is/are responsible?
Do a least significant difference calculation. (Section 4.5)
45. Why do I keep seeing an error message when I do a two-way ANOVA with replication in Excel?

You must choose all the data and the column and row headers too. Also make sure you have equal numbers of replicates for each instance of the factors. (Section 4.9)

46. Why should I bother plotting a calibration graph when I could simply use the regression equation?
    The plot serves as a good visual check for curvature that may still give a high r^2 or low $s_{y/x}$. Always plot the residuals too! (Sections 5.2, 5.3.2)
47. How many points should I have in my calibration?
    A minimum of six. (Section 5.6)
48. In the calibration equations sometimes the symbols for X and Y are upper case and sometimes they are lower case (x and y). When do you use upper case and when do you use lower case?
    Upper case letters are used for a quantity, for example Y may be the current at a glucose electrode. Small letters denote a particular quantity, for example $y = 10\,\text{nA}$. Example: a correct statement of a t-test is that $p(T \geq t) = 0.05$ which reads: the probability of finding a Student t value (T) equal to or greater than the t calculated from the data is 0.05. (Section 5.3)
49. In a calibration equation $y = a + bx$ what are the units of a (the intercept) and b (the slope)?
    a has the same units as y/x while b has the units of x. (Section 5.3.1)
50. When do you use a hat ($\hat{\ }$) on symbols and when a bar ($\bar{\ }$)?
    A hat (e.g., $\hat{x}$) indicates an estimated quantity. For example, in analysis this can be a result derived from a calibration procedure. A bar over a quantity denotes an average, for example $\bar{x}$ ($n = 4$). (Sections 5.3, 2.4.1)
51. I have used LINEST in Excel, but only get one value.
    You need to hold down Control-Shift while pressing Enter. If you accidentally press Enter and the output array is no longer selected, simply reselect the array and place the cursor in the command line again and hit Ctrl-Shift-Enter. Also make sure you have highlighted a block of cells 5 rows $\times$ 2 columns, and that the last (fourth) parameter is set to 1. (Section 5.4)
52. I have calibration data but how do I determine the uncertainty in my estimate of the unknown?

Use equation 5.15 and the relevant values from LINEST. (Sections 5.3.1, 5.4.3)

53. What is the best Excel function to estimate my regression equation and associated uncertainty?
    We recommend LINEST. (Section 5.4.3)
54. When trying to determine the detection limit, what if I cannot make a blank measurement?
    Make a series of measurements near the expected detection limit and use the calibration formula (equation 5.28). (Section 5.8)
55. Should I use r or r^2 to indicate the linearity of my calibration?
    (*Alternative*: Everybody I know uses R^2 as an estimate of the quality of a calibration equation. Is this okay?)
    Neither. These tell you about the linear relation between y and x, true, but in analytical chemistry you are rarely testing the linear model. The standard error of the regression ($s_{y/x}$) is a useful number to quote, or calculate 95% confidence intervals on parameters and estimated concentrations of test solutions. Plot residuals against concentration if you are concerned about curvature or heteroscedacity. (Sections 5.3.2, 5.5)
56. When are the degrees of freedom $n-1$ and when are they $n-2$ in an n-point calibration?
    Degrees of freedom $= n -$ the number of parameters calculated, so if you force the intercept to zero ($Y = bx$) then $df = n-1$, and if you calculate an intercept ($Y = bx + a$) then $df = n-2$. (Section 5.3.1)
57. How can I determine whether a point is an outlier in a calibration plot?
    As a rule of thumb, if the residual of a point has a magnitude 3 times greater than $s_{y/x}$ the point is suspect. (Section 5.3.2)

Some Useful Excel Functions

Remember that Excel does not know about units and always works at full precision. When you finally transcribe results into a report think about the appropriate units and significant figures.

How Do I...	*What to Do in Excel*
Quote a mean and sample standard deviation of data?	=AVERAGE(*range*) =STDEV(*range*) where the *range* is a list of cells that contain the data, e.g., A1:A20, B1:H1
Quote the % relative standard deviation of data?	=100* STDEV(*range*)/AVERAGE(*range*)
Quote the 95% confidence interval of the mean of n data?	=STDEV(*range*)* TINV(0.05, *n*−1)/SQRT(*n*)
Determine the probability of a t-value?	=TDIST(*t*, *df*, *tails*) *tails* = 1 (one-tailed) or 2 (two-tailed) *df* = degrees of freedom
Quote the median of data?	=MEDIAN(*range*)
Determine how many experiments I should do to ensure that my mean is within a certain tolerance of the true mean with 95% probability given the population standard deviation?	=ROUNDUP(NORMSINV(0.025)* σ/ε)^2,0) σ is the population standard deviation ε is the permissible tolerance
Calculate the interquartile range (IQR)?	=(QUARTILE(*range, 3*) – QUARTILE(*range, 1*))
And normalized IQR?	=(QUARTILE(*range, 3*) – QUARTILE(*range, 1*)) *0.75
Calculate a two-tailed Student t-value for a 95% confidence limit?	=TINV(0.05, *df*) *df* = degrees of freedom
Calculate a one-tailed Student t-value for a 95% confidence limit?	=TINV(0.025, *df*) *df* = degrees of freedom
Calculate the probability of a Student t-value?	=TDIST(*t*, *df*, *tails*) *df* = degrees of freedom *tails* = 1 (one-tailed) or 2 (two-tailed)
Calculate the critical G value, $G_{critical}$, at 95% probability for a Grubbs's outlier test?	=(n – 1)/SQRT(n)*SQRT((TINV(0.05/ n, n – 2))^2/(n – 2+TINV(0.05/ n, n – 2)^2))
Calculate a one-tailed Fisher F value at 95% probability?	=FINV(0.05, df_1, df_2) df_1 = degrees of freedom of the numerator df_2 = degrees of freedom of the denominator

Calculate the probability of an F-value?	=FDIST(F, df_1, df_2) df_1 = degrees of freedom of the numerator df_2 = degrees of freedom of the denominator
Fit a linear equation ($Y = a + bx$) to a set of x, y data, and calculate the standard error of the regression ($s_{y/x}$), and standard errors of slope (s_b) and intercept (s_a)?	b = SLOPE(*y-range*, *x-range*) s_b = INDEX(LINEST(*y-range*, *x-range*,1,1),2,2) a = INTERCEPT(*y-range*, *x-range*) s_a = INDEX(LINEST(*y-range*, *x-range*,1,1), 2,1) $s_{y/x}$ = INDEX(LINEST(*y-range*, *x-range*,1,1),3,2)
Calculate the standard error of the estimate of x ($s_{\hat{x}}$) from a measurement of y (y_0) and a linear calibration?	= $s_{y/x}$/b* SQRT(1/m + 1/n + (y_0 − *ybar*)^2/b^2 /SUMSQ(*xbar-range*)) $s_{y/x}$ is a cell containing the standard error of the regression (see previous entry) b is the slope of the calibration curve m is the number of repeats of the test solution (if $m > 1$, y_0 is the mean of the m replicates) n is the number of points in the calibration curve *ybar* is the mean of the calibration y values *xbar-range* is a range containing the x calibration values minus the mean of the x calibration values (mean centered x values)
Calculate 95% confidence interval on slope, intercept, and estimate of x?	=TINV(0.05,n − 2)*s_b =TINV(0.05,n − 2)*s_a =TINV(0.05,n − 2)*$s_{\hat{x}}$
Draw the best-fit line through the experimental points graphed in an X–Y (scatter) chart? (Note: it looks better if your chart starts with just the data points with no connecting lines.)	1. Right click on a point in the chart. Left click on Add Trendline... 2. Click OK
Calculate the estimated y values in a calibration?	= TREND($*y-range*, $*x-range*, *x*, *inter*) where *inter* = 1 for an intercept and 0 to force the line through zero. Note the $ before the x and y ranges (i.e., write as $A$1:$A$10). When you copy the formula down for all the x values, you only want the particular x to change, not the ranges for the calibration!

1

Introduction

1.1 What This Chapter Should Teach You

- To understand that chemical measurements are made for a purpose, usually to answer a nonchemical question.
- To define measurement and related terms.
- To understand types of error and how they are estimated.
- What makes a valid analytical measurement.

1.2 Measurement

Chemistry, like all sciences, relies on measurement, yet a poll of our students and colleagues showed that few could even start to give a reasonable explanation of "measurement." Reading textbooks on data analysis revealed that this most basic act of science is rarely defined. Believe it or not there are people that specialize in the science of measurement: a field of study called metrology. The definition used in this book for measurement is a "set of operations having the object of determining the value of a quantity." We will come back to this but first...

1.3 Why Measure?

The world spent an estimated US$3.1 billion on chemical measurements for medical diagnosis in 1998, most of this measurement being done in the United States and the European Union. These measurements were carried out to discover something about the patients.

The sequence of events that involve a chemical measurement are: (1) state the real-world problem; (2) decide what chemical measurement can help answer that problem; (3) find a method that will deliver the appropriate measurement; (4) do the measurement and obtain a result (value and uncertainty, including appropriate units); and (5) give a solution to the problem based on the measurement result. It is important to understand the relationship between the real-world problem and the proposed measurement. The chemical measurement may give only part of the answer, and should not be confused with the answer itself. In forensic analytical chemistry, matching a suspect's DNA with DNA sampled at the crime scene does not necessarily mean that the suspect is guilty. In health care, a cholesterol measurement might tell the doctor about the likelihood of a patient contracting heart diesease, but a full analysis of high- and low-density lipids and other fats will be more useful.

1.4 Definitions

Our definition of measurement as a "set of operations having the object of determining the value of a quantity" comes from the bible of metrology (the International Vocabulary of Metrology or the VIM). To really understand this definition we need to know what a quantity is. A quantity is defined as "attribute of a phenomenon, body, or substance that may be distinguished qualitatively and determined quantitatively." Think of things that you measure and see how they fit into this definition. As chemists we often measure the concentration of a particular compound. The substance of which we wish to know the concentration must be stated (= distinguished qualitatively). This is obvious for many measurements (e.g., the concentration of sodium chloride in seawater) but may be less so when issues of isomerization (D- or L-thalidomide, or both), or speciation (chromium(VI) or total chromium), or more nebulous definitions (pH 8 extractable organics) arise. Determining the value of a phenomenon may refer to activities such as measuring the rate constant of a reaction, or the amount of solar energy falling on the Earth. Finally (before we can get stuck into measurement) we need to know what a "value" is. A value is the "magnitude of a particular quantity generally expressed as a unit of measurement multiplied by a number."

1.5 Calibration and Traceability

Measurement is, therefore, something we do that results in a number and a unit. How we obtain that number is the point of the experiment, but it usually involves comparing our unknown system with a known system, either directly, as happens when we measure the length of something using a ruler suitably marked in length units, or indirectly, as in when we calibrate an instrument and then measure the sample for analysis. Indirect comparisons are often made in modern chemistry. Peaks in a gas chromatogram of a test material of unknown concentration may be compared with those from a series of materials of known concentrations via a linear calibration graph to obtain the value of the unknown. In the case of blood glucose concentration the instrument response is an electric current that is proportional to the concentration of glucose. The proportionality constant for monitors used by patients in their homes is established in the factory so that their monitor reads the glucose concentration directly. The unit of the measurement is taken care of by knowledge of the units of the quantities of the known samples. The measurement of the concentration in the gas chromatography example is the determination of the peak height of the sample plus the calibration followed by an appropriate calculation. It is important to realize, too, that instruments that appear to give us the answer directly in the necessary units, for example a pH meter, are only doing so courtesy of electronics that can compare electrical signals (in the example, potential measurements of a glass electrode) from the application of the instrument to a known standard (a calibration buffer solution) with those from the unknown sample. In chapter 5 we show how calibrations can be established and used to deliver values of the measurand.

1.6 So Why Do We Need to Do Data Analysis At All?

The need for data analysis in any measurement science is a consequence of measurement uncertainty. Having made our measurement, and before we try to interpret the result, an immediate question is, or should be, "How reliable is the result?" The nonscientific public is used to accepting measurements at face value. We rarely question the weight of baked beans written on the outside of a supermarket can, or of potatoes indicated by the scales at the checkout. In courts, drivers usually accept the evidence of the police radar that they were speeding, or the breathalyzers that indicated they were over the limit of blood alcohol. However, the prospect of a loss of license or even time in jail has caused some defendants to try to challenge those measurements. In trade, when small differences in a measurement result, say the protein content of wheat, can lead to thousands of dollars more or less to buyer or seller, measurements can frequently be scrutinized and argued over. When it matters, we become keenly aware of the importance of accurate measurements. Any chemical measurement that is worth doing is of importance to someone and the modern analytical chemist must give information of the reliability of the result. In fact, any analysis without proper information of the reliability is useless!

Modern analytical chemists may not understand how far-sighted the Swedish chemist Berzelius was when he wrote, in the 19th century, concerning the mission of the analyst "not to obtain results that are absolutely exact—which I consider only to be obtained by accident—but to approach as near accuracy as chemical analysis can go." No amount of modern nano-machines, spectrometers, or expensive instruments will overcome this statement of a universal truth. We can minimize the uncertainties associated with measurement. We can estimate the uncertainties, but the "absolutely exact" results lie permanently beyond our grasp.

The purpose of this book is to furnish you with tools to help you maximize the quality of your results; that is, when you, a chemical analyst, give a result it is the best possible and is accompanied by a true statement of its reliability.

1.7 Three Types of Error

Here we discuss the concepts of "error" and "uncertainty." In the world the word "error" implies a failure of some kind—synonyms include "mistake," "blunder," "slip," and "lapse." In metrology, error is defined as "the result of a measurement minus a true value of the measurand" and is free of such negative connotations. Error in an analysis is a particular value that may be known if the true value is given.

If we conduct an experiment we almost always obtain a result that is in error. Why did we not get it right? We could have simply made a mistake in weighing, calibration, or even the calculation. Repeating the experiment might show up this error. The first type of error where we make a mistake, *gross error*, is really a fault and neither this nor any book on analytical chemistry can help. It may be possible to identify such an error, perhaps by statistical analysis, and remove that result from further consideration, but there is no other way we can usefully employ the result. If justified statistically (we'll come to this in chapter 3) it should be taken out, after, of course, carefully noting the fact in a laboratory notebook. Please note that careless and unrecorded expunging of results could amount to scientific fraud. Beware the outlier that turns out to be the only halfway decent result! However, whatever the fate of this grossly erroneous result, because of its unique nature, it cannot guide our future actions.

With regards to why we did not get our analysis completely right, the second possibility is the method itself may be flawed. No amount of repeats will improve the situation. This second type of error, *systematic error*, is a permanent deviation from the true result. When applied to an instrument, systematic error is known as *bias*. A colorblind person might persistently overestimate the end point in a titration, the extraction of an analyte from a sample may only be 90% efficient, or the derivatization step before analysis by gas chromatography may not be complete. In each of these cases, if the results were not corrected for the problems, they would always be wrong, and always wrong by the same amount for a particular experiment. Systematic error can be estimated by measuring a reference material a large number of times. The difference between the average of the measurements and the value of the reference material is the systematic error. It is always desirable to know the sources of systematic error in an experiment and to correct for them in measurements.

In the description of how to estimate a systematic error, it was suggested that the experiment be repeated a large number of times. This is necessary because of the contribution of another source of error, namely *random error*. Random error is the third type of error that could be responsible for why the answer in our experiment is in error. Despite your best efforts, having considered and removed or corrected for sources of systematic error, having ironed out gross errors, repeating experiments always seems to give slightly different answers. Sometimes the result is a bit more than expected, sometimes

Table 1.1 Types of error

Gross	Blunders
Systematic	Always the same value and sign
Random	Normally distributed with mean of zero

a bit less. Rarely it appears to be a long way from the accepted value. The good news is that taking the average of a large number of results seems to give an acceptable answer. The values that are too high cancel those that are too low. There are a myriad of factors that can contribute to random error: the inability of the analyst to exactly reproduce conditions, fluctuations in the environment (temperature, pressure), rounding of arithmetic calculations, brief gusts of wind, or a shake of the analyst's hand. What do not contribute to random error are changes in conditions such as the regular drift in baseline of an instrument and the aging of a chromatography column.

Table 1.1 summarizes the three types of error.

1.7.1 An example—pipetting

Considering why we might not deliver exactly 10 mL using a 10 mL pipette is instructive (figure 1.1).

We shall identify three contributing factors to the problem. There are more, and as an exercise the reader might try to think of everything that can go wrong with this apparently straightforward operation in analytical chemistry.

1. The manufacturer will admit that the pipette you are using, when filled properly to the mark at 20°C, is only guaranteed to have a volume somewhere between 9.98 and 10.02 mL. Perhaps you are lucky and have a 10.00 mL pipette, but perhaps not. Note that any error of this type is a systematic error.
2. When you use a pipette, do you really fill it exactly to the same mark each time? A series of 10 experiments of filling a pipette with distilled water and weighing what runs out, gives a range of values from 9.95 to 10.04 mL. The analyst's contribution to the error is definitely random.
3. You are aware that during your experiments the temperature in your laboratory fluctuates between 19.2 and 23.1°C, and

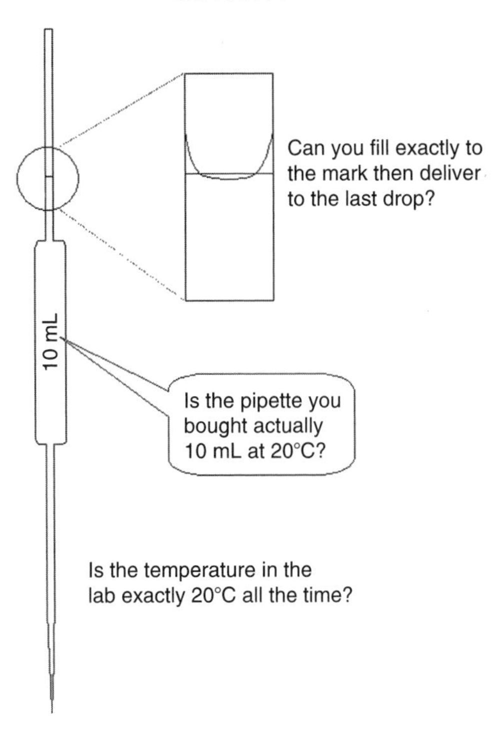

Figure 1.1 Uncertainties and errors in delivering 10 mL by a pipette.

you know that the volume of 10 mL of water will increase by 0.0021 mL for every degree centigrade rise in temperature. If the experiments take long enough to allow the temperature to change in a random fashion about some average, then these changes will be included in the distribution of results from the fill-and-weigh experiments in 2. In addition, unless the average temperature during the experiments was exactly 20°C there will also be a systematic error arising from the difference.

1.7.2 An example—the Royal Australian Chemical Institute titration competition

Although the above sounds plausible, do we have any evidence for these definitions of "error"? Take as an example the Royal Australian Chemical Institute's (RACI) schools titration competition of 1997. In this competition, each of a team of three high school students is asked

Table 1.2 The results of the 1997 RACI titration competition. The values are independent students' results for the concentration of a solution of acetic acid (units: M). The correct answer was 0.1147 M

0.1150, 0.1152, 0.1143, 0.1144, 0.1153, 0.1138, 0.1139, 0.1150, 0.0920, 0.1556, 0.1141, 0.1219, 0.1222, 0.1143, 0.9083, 0.1134, 0.0936, 0.1155, 0.1145, 0.1177, 0.1146, 0.1158, 0.1142, 0.1148, 0.1144

to measure the concentration of a solution of acetic acid, given a solution of sodium hydroxide and a solution of hydrochloric acid of known concentration.

The members of the team use the hydrochloric acid solution to standardize the sodium hydroxide solution, which in turn is used to titrate the acetic acid solution. In table 1.2 are the results of one member of each of 25 teams that participated in the competition in 1997 at the University of New South Wales, Sydney.

Although no one is spot on, there are some very near misses and some pretty woeful answers (e.g., 0.9083 M). We know that the students were not given enough sodium hydroxide to titrate 25 mL of 0.9083 M acetic acid, so we can confidently say that this is a gross error in calculation or recording the result. Plotting the results in increasing order of magnitude reveals some interesting groupings (figure 1.2).

The high result is clearly off the planet, as can be seen from figure 1.2, but we will show that six other results can also be classed as outliers by methods we explain in chapter 3. The remaining data are shown to group around the accepted answer, and within limits of an expected random scatter, calculated from an analysis like the one above for the pipette, but covering all sources of uncertainty. Out of the 25 results, seven are rejected as gross errors and of the remainder seven fall above and 11 fall below the correct answer. Plotting a histogram, a bar chart of how many results fall within given ranges, reveals the distinction between the random error and the outliers (figure 1.3).

The two bars at each end represent the numbers of students whose values were more than 2.5% away from the correct result. Perhaps surprisingly, there are hardly any who are 2% away. It seems if you are going to stuff up, you will do it big time, and the rest are going to be distributed about more or less the right answer. The peak of the histogram is just on the high side of 0% error, but looking at the

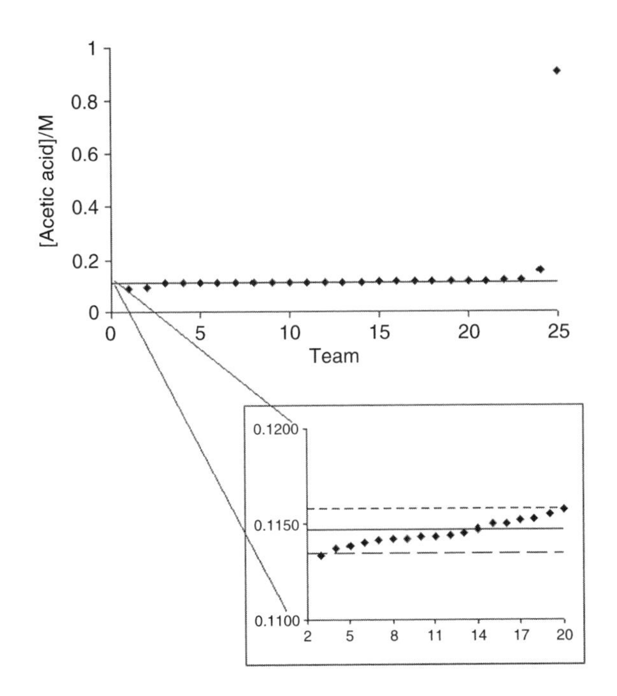

Figure 1.2 Results of the 1997 RACI titration competition. Inset: results for teams 3–20. The line is the accepted result (0.1147 M) and the dashed lines are ±1%.

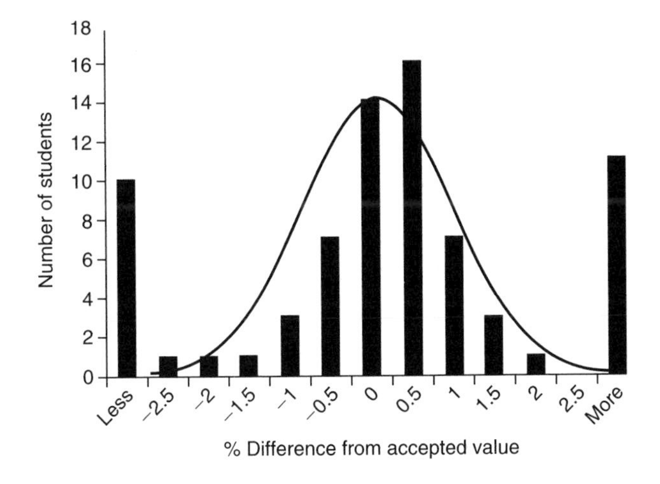

Figure 1.3 Histogram of the 1997 RACI titration results. Each bar is the number of students whose result fell between the number indicated and the number to the right. Note that the 25 data points in table 1.2 represent a subset of all the data from the RACI titration competition. The entire data set of 75 results was used to generate this histogram.

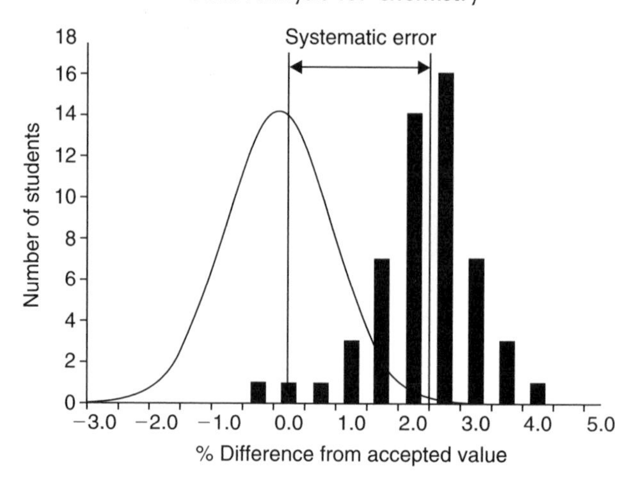

Figure 1.4 Histogram of the 1997 RACI titration results if there had been a systematic error.

spread of results, there is no evidence to suggest a significant systematic error. We shall see later that the bell-shaped curve overlaid on the histogram represents an idealized spread of the data.

In a hypothetical example, if the peak of the histogram were at, say, +2.5% (see figure 1.4), we would conclude that there was a positive systematic error in the measurement. In this hypothetical case perhaps the RACI had got it wrong and the actual concentrations of the acetic acid solutions were more than they thought, or perhaps the concentration of the standard hydrochloric acid was less than given.

Data analysis might reveal what has happened, but it will not tell you why it happened.

Why have we gone to the trouble of classifying different types of error? Because once we can identify the systemic errors we can correct for them, and a statistical treatment of the random error will allow us to estimate what the true result is and what uncertainty there may be about that result. Figure 1.5 brings together this discussion and shows the relationships between the true value of the measurand, the errors in a single measurement result, and the distribution of random errors.

1.7.3 Measurement uncertainty

The discussion of errors given above is known as the "classical approach" to measurement. It has served measurement science well,

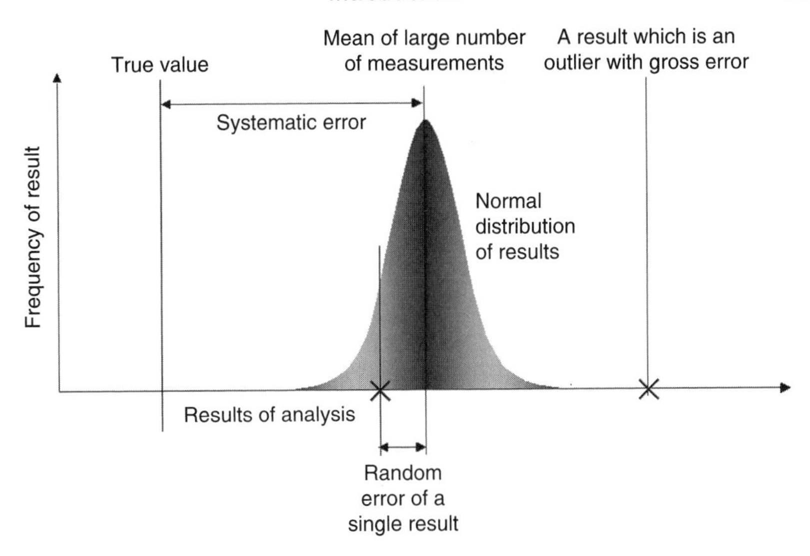

Figure 1.5 Errors in measurement results, showing the difference between systematic, random, and gross errors.

but it is based on the assumption that a measurand can ultimately be described by a single true value. There has been a shift in recent years to an understanding that the concept of a "true" value may not be correct, and therefore notions of accuracy and random and systematic errors may also not be valid. The "uncertainty approach" understands there is only one uncertainty of measurement, ensuing from various components. It characterizes the extent to which the unknown value of the measurand is known after measurement, taking account of the given information from the measurement. Some of the statistical tools we explain in this book are necessary for the estimation of this uncertainty, and the classical approach is a useful starting point for a discussion of the nature of chemical measurement.

1.8 Accuracy and Precision

Accuracy is a concept that encompasses getting the answer right (sometimes known as trueness) with acceptable uncertainty (i.e., with good *precision*). The relationship between accuracy and precision is shown in figure 1.6 where high precision is represented by the closeness of the cluster of hits on a target and high accuracy is

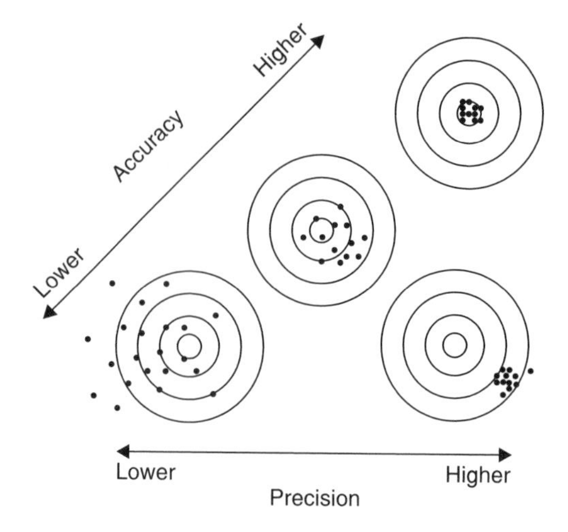

Figure 1.6 Relationship between accuracy and precision.

represented by the hits being centered around the bull's-eye. In the past accuracy just meant getting the answer right, but it is now understood that trueness without precision is not a desirable state. Accidentally achieving a reasonable answer but with a huge uncertainty still leaves the user of the chemical information unsure as to the validity of the result. The axes are not orthogonal as these concepts are not entirely independent.

1.8.1 Were the RACI students accurate or precise?

Were the RACI titrators accurate or precise? Will they make good analytical chemists? Apart from the poor souls who wasted their time and managed to obtain outrageous answers, the majority did very well indeed. They averaged out at the correct answer, and although the spread of results was a bit greater than might be expected from the best analytical practice, the school students performed well. If we needed to know what the concentration of acetic acid was, then the results given by the majority of students could be called accurate.

1.8.2 How to estimate precision

We have discussed uncertainty of a measurement result in terms of a possible spread of values. In the RACI competition it appears that students whose experiments were subject to random error find results that spread about 2% either side of the mean value. More achieve values nearer the mean, and the results are distributed evenly about the mean, with about half falling greater than the mean and half falling less than the mean. Many analytical results when repeated show exactly this pattern. It can be shown that data that follows this random distribution about a mean can be described by a normal, or Gaussian, probability density function (pdf)

$$f(x|\mu,\sigma) = \frac{1}{\sigma\sqrt{2\pi}}\exp\left[-\frac{(x-\mu)^2}{2\sigma^2}\right] \quad (1.1)$$

Note that the pdf is a function of x—the values that can be taken by the data. A probability density function is defined in terms of its area; the probability of finding a result between two values of x (say x_1 and x_2) is the area under the pdf between x_1 and x_2. The shape of this pdf is the familiar "bell-shaped curve" shown overlaying the histogram of figure 1.3.

The pdf is characterized by two parameters: μ the mean for the infinite population of data that define the pdf, and σ^2 the variance (we discuss the mean and variance in more detail in chapter 2). The maximum of the function is when $x=\mu$, and the larger the value of σ, the more spread out the function is. If the data were results of repeated analyses of a sample, then σ is a measure of the precision of the analysis. Equation 1.1 is not the probability of finding a particular result, but it is related to it. The integral of f with respect to x between limits $x=a$ and $x=b$ is the probability of finding a result in that range. The curve in figure 1.3 is generated from equation 1.1 with $\mu=0.069\%$ and $\sigma=0.84\%$ (remember we are determining the distribution of the error of each titration result).

The square root of the variance is called the standard deviation. Although knowledge of the exact form of equation 1.1 may not be of great interest to a chemist, it is useful to know something of its properties. It turns out that for all normal distributions the area under the curve from $\mu-\sigma$ to $\mu+\sigma$ is 68% of that from $-\infty$ to $+\infty$. If the range is widened to $\pm\ 2\sigma$, then just over 95% of the entire area is covered, and for $\pm\ 3\sigma$ 99.7%. Figure 1.7 shows limits that enclose different percentages of the normal distribution. Remember that to reach 100% you have to go a long way ($\pm$ infinity!).

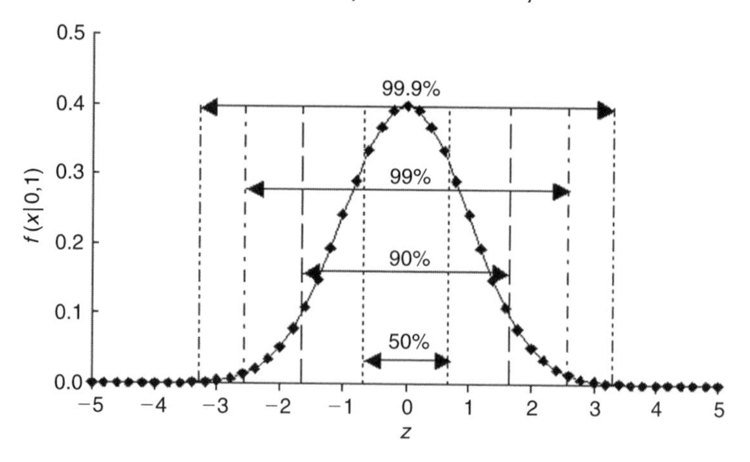

Figure 1.7 Probability distribution function (pdf) of the normal distribution with $\mu = 0$ and $\sigma = 1$. z is the number of standard deviations from the mean. The ranges shown contain the percentages of the distribution given.

The fortuitous coincidence of a nearly integral number of standard deviations ($\pm 2\sigma$, more precisely $\pm 1.96\sigma$) covering a nice fraction of the distribution (95%) has probably led to the popularity of 95% confidence intervals, and tests at 95% probability. Remember that the area under the curve is the integral, the fact that $\pm 2\sigma$ encompasses about 95% of the area also means the probability of finding a value between the limits of $\pm 2\sigma \approx 95\%$. Imagine the RACI competition was extended to every school student in the world who is doing chemistry, and the mean and standard deviation were as reported above. Then we could say that if we picked 100 school students at random, checked to make sure none of them had made a gross error, 95 of them would be expected to have results within the range $\mu \pm 2\sigma$ (between $0.069 - 0.84 \times 2 = -1.61\%$ of the given concentration and $0.069 + 0.84 \times 2 = 1.75\%$ of the given concentration). This may be very useful. If the mean and standard deviation are known, and that the population is normally distributed, then we can say many things about likely results without ever doing another experiment. We might use our knowledge of this distribution to decide which students have performed the analysis badly, and which have done well and whose results are bona fide members of the distribution of correct results. A student who is out by 5% has a probability of doing so by purely random processes of one in one billion ($1{:}10^9$). It may be that we are more willing to believe that he or she committed a gross error than just happened to be a very unlikely member of the normal population.

We shall see in the next chapter that this analysis implies knowledge of μ and σ, and these statistics are not always available, and are certainly not the same as the mean and sample standard deviation of a small data set.

1.9 Significant Figures

The way a result is written should tell something about the precision of the result. The more figures quoted the greater the implied precision. The significant figures are those that impart useful information. It would not be at all appropriate to quote the length of a swimming pool to a fraction of a millimeter.

1.9.1 Counting significant figures

If it is not obvious then the number of significant figures is best determined by writing the result in scientific notation (i.e., $x.xxx \times 10^y$) and counting the digits.

Example 1.1

State how many significant figures there are in the following measured amounts.

1. The concentration of copper in tap water was 0.00000572 M
2. The concentration of glucose in blood was 5.0 mM
3. The mass of ammonium nitrate was 5.20 tonnes

Solution

1. Expressing the concentration of copper in scientific notation, 0.00000572 becomes 5.72×10^{-6} M. Hence there are three significant figures, the digit before the decimal point and the two digits after the decimal point.
2. The 5.0 mM concentration of blood can be re-expressed as 5.0×10^{-3} M and therefore the number of significant figures is two.

3. The 5.20 tonnes of ammonium nitrate is equivalent to 5200 kg. Expressed in scientific notation the mass is 5.20×10^3 kg and there are three significant figures.

Comment

1. In the copper concentration of 0.00000572 M the zeros before the 5 do not count as significant figures because the zeros are only being used to locate the decimal point. If this were not the case, simply by changing our units we could alter the number of significant figures.
2. In the blood glucose example the zero is counted as it is listed in the original value, which suggests we know the glucose concentration is 5.0 not 5.1 or 4.9. However, we do not know whether 5.0 is really 5.01 or somewhere between 4.95 and 5.04.
3. Note with the mass of ammonium nitrate being 5200 kg we have no idea whether either of the two zeros is significant or not and therefore we would make the assumption that neither is significant and we would conclude there are only two significant figures. The expression of the mass as 5.20 tonnes, however, tells us that the first zero is significant as it is included in the scientific notation and that the second is not significant. The use of units with prefixes such as m or μ is a way of applying scientific notation.

1.9.2 How many significant figures?

For a measurement made by a modern instrument the figures are usually output digitally and should be used as given even if it is not clear that all are really significant (see the gas chromatography example 2.1a). It is only when finally writing the result that the number should be rounded to an appropriate number of significant figures. Let the spreadsheet do all the calculations and leave any concerns about significant figures until the end when the final result is required.

To decide what the correct number is, it is necessary to know about the uncertainty of the measurement. This may be derived from the analyst's knowledge, experience, or common sense, or may be determined from the standard deviation of repeated experiments.

For example, the use of a 30 cm ruler, graduated in millimeters, to measure a length of about 10 cm, will be reasonably given to the nearest millimeter or half millimeter (e.g., 10.3 cm or 10.25 cm).

If the standard deviation of a result is known, or if a 95% confidence interval has been calculated, this is a guide to the number of significant figures.

- Write the standard deviation or uncertainty to two significant figures.
- Then write the result to the same order of magnitude (i.e., powers of 10, or decimal places).

Therefore, if the concentration of acetic acid in the titration competition was determined as 0.1146 M with a 95% confidence interval of 0.0096 M then we could state the value of the concentration of acetic acid as 0.1146 ± 0.0096 M (95% confidence interval). However if the uncertainty were 0.011 M then the concentration would be expressed as 0.115 ± 0.011 M (95% confidence interval).

In the examples given throughout this book we will highlight the answers for which the number of significant figures have been determined by this rule.

1.10 Fit for Purpose

Ultimately, the results determined by an analytical chemist have to be good enough, accurate enough, to allow the proper use of them. The concept of "fit for purpose" sums up what is required. Remember no one wants analytical chemistry for its own sake. They want to know if they can eat the food, drink the water, invest in the gold mine. The quality of the analytical chemistry needs to be sufficient to answer the question. A litmus paper test for pH could well have an uncertainty of 2 pH units, but if the interest is only to find out if the solution is acidic, then litmus paper is entirely fit for purpose, and the use of a carefully calibrated pH meter would be overkill.

The United Kingdom Laboratory of the Government Chemist has proposed six principles of valid analytical measurement (VAM):

- Work to an agreed customer requirement.
- Use validated methods and equipment.

- Use qualified and competent staff.
- Participate in independent assessment of technical performance (proficiency testing).
- Ensure comparability with measurements made in other laboratories (traceability and measurement uncertainty).
- Use well-defined quality control and quality assurance practices.

The methods of data analysis described in this book will be of use in fulfilling each of these principles. Without a proper understanding of the statistics of data an analyst cannot hope to deliver results that are "fit for purpose."

2

Describing Data: Means and Confidence Intervals

2.1 What This Chapter Should Teach You

- To understand the concept of mean, variance, and standard deviation pertaining both to a large sample (population) and a small sample.
- To define the standard deviation of the mean of a number of repeated measurements and understand its relation to the sample standard deviation.
- To define confidence intervals about a mean and show how to use them to indicate measurement precision.
- To introduce robust estimators of representing the average and sample standard deviation.
- To appreciate the difference between measurement repeatability and reproducibility.

2.2 The Analytical Result

Why do we bother with means and standard deviations? Because these two statistics tell us a great deal about the data and the population from which they come. A mean of a number of repeated measurements of the concentration of a test solution is an estimate of the concentration of the test solution and the sample standard deviation gives a measure of the random scatter of the values obtained by measurement. Together with the appropriate units they represent the result. This information is not necessarily the answer to: "What is the concentration of the test solution and how sure are you of that answer?" To answer this question an uncertainty budget must be prepared, which includes errors, random and systematic, arising from all aspects of the experiment (of which the standard deviation of repeated measurements is just one).

Why is it good to repeat analytical measurements? There might be an argument for the "quit while you are ahead" school but repeating a measurement gives increased confidence in the result, especially if the numbers appear to agree. But apart from the appearance of consistency, do you get better answers by repeating measurements, and are more repeats better than fewer repeats? The answer to both questions is "yes," as we shall see in this chapter. Note that the statistical treatment of repeated results does not tell us about systematic error unless we can compare our mean with a known or assigned value of the quantity being measured.

2.3 Population and Sample

A statistician calls the infinite number of results that could be obtained and that are described by the probability distribution function (see chapter 1) the *population*. The distribution of the values of those results is characterized by the population mean μ and the population standard deviation σ. The goal of many of our data analysis methods is to estimate μ and σ from only a few repeated measurements called a *sample*. (In this respect the definition of population differs from the biologist's view of a finite population of organisms.)

There is a small problem here experienced only by chemists. We tend to call each thing that we analyze a "sample," so have a firm idea that we have n samples that give our n data. Statistics refers to the n data as the sample to distinguish it from the infinite population. The International Union of Pure and Applied Chemistry (IUPAC) has recommended that the "actual material being studied" should be a description of the material preceded by the word "test," for example "test solution" or "test extract," and the word "sample" be reserved for its statistical sense. Although we shall try to adhere to this convention in this book, many scientists will find it almost impossible to not refer to a test solution as a sample.

2.4 Mean, Variance, and Standard Deviation

The concepts of mean (average), variance, and standard deviation have been introduced in chapter 1. Here they will be defined.

2.4.1 Mean

The sample mean (arithmetic mean, average) is the result of summing all the results and dividing by the number of data (n):

$$\bar{x} = \frac{\sum_{i=1}^{i=n} x_i}{n} \tag{2.1}$$

The usual symbol for the sample mean is the lower case symbol for the quantity (here x) with a bar across the top.

For normally distributed data the sample mean $\bar{x}$ tends to the population mean μ as the number of data becomes great. This is good, because the population mean is the true result in the absence of systematic error. Although a single result taken from a normally distributed population is more likely to be nearer the mean than farther away, in the absence of any other information about the population we have no idea how near that one result is. To illustrate the effect of taking means of increasingly larger samples, consider figure 2.1. We have generated 500 random numbers with mean $\mu = 10$ and standard deviation $\sigma = 1$, and have averaged $2, 3, 4, \ldots 500$ of them. Figure 2.1 is a plot of the means of n values against n.

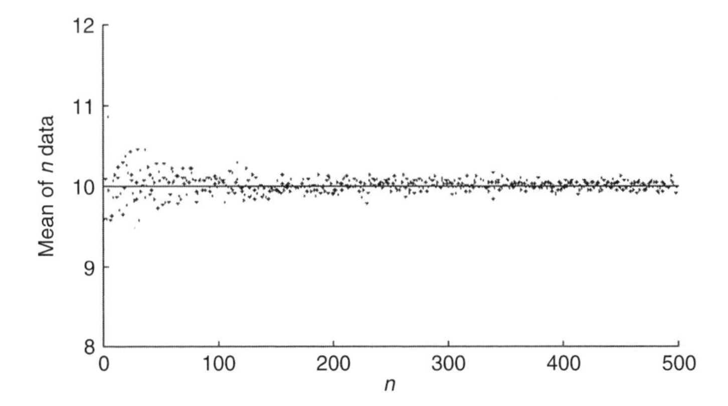

Figure 2.1 Means of n data randomly drawn from a normally distributed population with $\mu = 10$ and $\sigma = 1$ as a function of n.

Indeed, although it wobbles about to begin with, the means of the data do seem to fall nearer to the population mean (here 10) as n increases.

This sounds about right, as we know that the data is symmetrically distributed about the mean, so we would expect that the results that were on the high side of the mean to cancel those on the low side, and the more results we had, the better the canceling. This answers our question as to why do repeats. The more repeats that are done, the smaller the uncertainty about the sample mean, and hence the more confident one becomes that the sample mean is a good estimate of the population mean.

2.4.2 Standard deviation and variance

The spread of the population shown by the fatness of the bell-shaped curve is measured by the standard deviation or its square, the variance. They are defined for a sample of n data by:

$$s = \sqrt{\frac{\sum_{i=1}^{i=n}(x_i - \overline{x})^2}{n-1}}$$
$$s^2 = \frac{\sum_{i=1}^{i=n}(x_i - \overline{x})^2}{n-1} \qquad (2.2)$$

The standard deviation defined in equation 2.2 is known fully as the *sample* standard deviation because it refers to a sample of n and is an estimate of the population standard deviation σ. Figure 2.2 shows

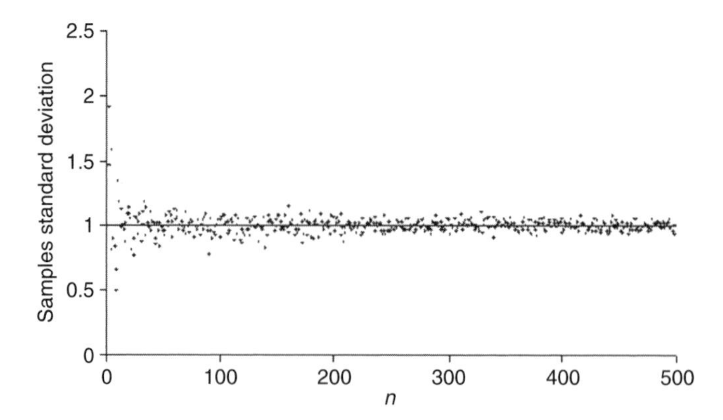

Figure 2.2 Sample standard deviations of n data randomly drawn from a normally distributed population with $\mu = 10$ and $\sigma = 1$ as a function of n.

how the sample standard deviation also converges on the population value (here $\sigma = 1$) as n increases.

2.4.3 Relative standard deviation

The relative standard deviation (RSD), also known as the coefficient of variation (CV), is the standard deviation of a measurement expressed as a fraction, or more usually as a percentage, of the mean:

$$\mathrm{RSD} = \frac{s}{\overline{x}} \times 100\% \tag{2.3}$$

The RSD of an analytical result is often quoted as it gives an immediate impression of the precision of the measurement. Less than 1% is usually considered very good for routine measurements which are more often in the 1–5% range. For the RACI titration competition (see chapter 1) the mean and standard deviation were 0.1146 M and 0.0006 M, respectively, after removing outliers from consideration. The RSD was therefore $0.0006/0.1146 \times 100\% = 0.5\%$, a very good result.

2.4.4 Units

Remember that the standard deviation has the same units as the mean, and therefore the variance has the units of the $(\text{mean})^2$. The relative standard deviation is a fraction with the same units for numerator and denominator, and therefore is unitless.

Example 2.1a

Calculation of the mean ($\overline{x}$), sample standard deviation (s), and relative standard deviation (RSD).

In an analysis to determine the ethanol content of a wine by gas chromatography, an internal standard of isopropanol is used to account for the variability in the volume injected between tests. In the measurement of a four-point calibration curve and the repeated analysis of the wine sample, six injections in all are performed. Each injection contained 1% v/v of the internal standard. The isopropanol peak area, in arbitrary units, for each of the six injections were 2957398, 3733127, 2900811, 3010190, 2810196, 2084063.

Problem

Calculate the injection precision (i.e., the standard deviation of the measurements).

Solution

To do this requires a determination of the mean, standard deviation, and RSD.

Calculation by Hand

The mean is calculated using equation 2.1:

$$\overline{x} = \frac{\sum_{i=1}^{i=n} x_i}{n}$$

$$= \left(\frac{\left\{ \begin{array}{l} 2957398 + 3733127 + 2900811 + 3010190 \\ + 2810196 + 2084063 \end{array} \right\}}{6} \right)$$

$$\overline{x} = 2915964$$

The standard deviation is calculated using equation 2.2:

$$s = \sqrt{\frac{\sum_{i=1}^{i=n} (x_i - \overline{x})^2}{n-1}}$$

$$= \sqrt{\frac{\left\{ \begin{array}{l} (2957398 - 2915964)^2 + (3733127 - 2915964)^2 \\ + \cdots + (2084063 - 2915964)^2 \end{array} \right\}}{5}}$$

$$s = 525705$$

The relative standard deviation is calculated using equation 2.3:

$$\text{RSD} = \frac{s}{\bar{x}} \times 100\% = \frac{525705}{2915964} \times 100 = 18.03\%$$

Calculation using Excel

1. Input the data into a column (here A2:A7)
2. To calculate the average, in a blank cell (A9) type in =AVERAGE(A2:A7)
3. To calculate the standard deviation, in a blank cell (A10) type in =STDEV(A2:A7)
4. To calculate the relative standard deviation, for spreadsheet 2.1 in a blank cell (A11) type in =(A10/A9)*100

Spreadsheet 2.1

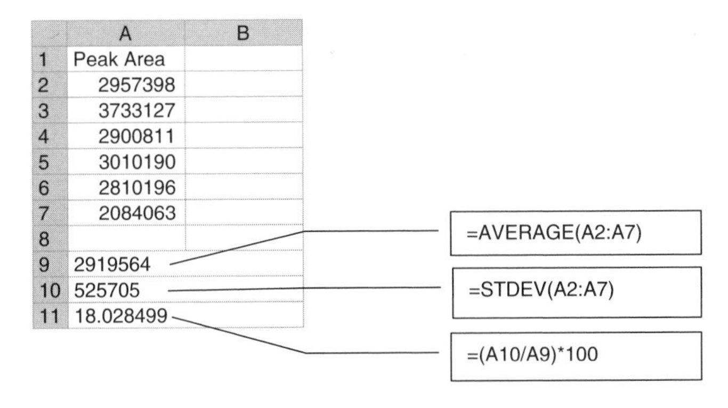

	A	B
1	Peak Area	
2	2957398	
3	3733127	
4	2900811	
5	3010190	
6	2810196	
7	2084063	
8		
9	2919564	
10	525705	
11	18.028499	

Answer

The mean of the isopropanol peak area is 2,920,000 with a standard deviation of 530,000 ($n=6$). The relative standard deviation of six injections of isopropanol is 18%.

Comments

1. Note the way the answer is expressed which informs the reader how many samples were analyzed.
2. Also note that the ± was not used as we recommend this is reserved for confidence intervals (see section 2.5).

3. Rather than calculate from first principles it would be more common to use a calculator or Excel. With a calculator there are often keys for population standard deviations (written as σ or $x\sigma_n$) and sample standard deviations (written as s or $x\sigma_{n-1}$). Naturally $x\sigma_{n-1}$ is the correct key to use for this example and gives a value of 525705. If $x\sigma_n$ was used you would get a standard deviation of 479900. As appealing as using a smaller value is, it is not correct to quote the standard deviation calculated in this way.
4. Remember significant figures. Do not round until you need to report the result. Then give the standard deviation or confidence interval to two significant figures and the mean to the same order of magnitude.

2.4.5 Degrees of freedom

It is seen from equation 2.2 that the more spread the data is about the mean, the bigger the standard deviation. A consequence of the square is that whether a datum is more or less than the mean its difference from the mean will always contribute positively to the standard deviation. The sum of the squares is divided by $n-1$, which is the number of degrees of freedom (*df*) of the calculation. Each data point gives one degree of freedom, and by calculating the mean to use in equation 2.2, one degree of freedom has been used up. Degrees of freedom appears in many equations and is defined as the number of data minus the number of parameters calculated from them. For calculation of many statistical parameters of sets of data $df = n-1$. The only times this will be different in this book are for calibrations for which two parameters, slope and intercept, are calculated, giving $df = n-2$.

2.4.6 Standard deviation of the mean

Just as we can see in figure 2.1 that the mean draws ever nearer to μ with increasing numbers of data, it must also follow that our confidence in that mean as an estimate of μ also increases. As an expression of this confidence it is possible to define a "standard deviation of the mean." To explain this, suppose we average four points, then another four, then another four, etc. The means of four points themselves form a group (in the jargon another "population")

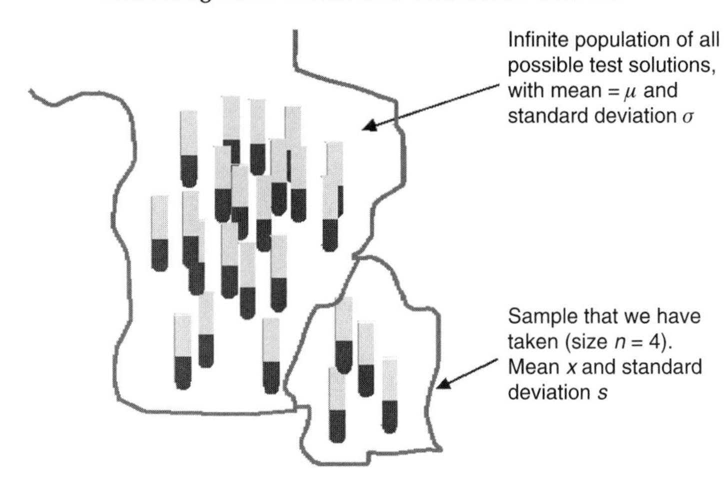

Figure 2.3 Taking samples of four test solutions from the population (hypothetical) of all possible test solutions.

with its own mean and standard deviation, which can be related to the mean and standard deviation of the original set of data from which the samples were drawn (see figure 2.3).

By a theorem in statistics called the central limit theorem, the mean of the means is, not surprisingly, the mean of the population, and the standard deviation of the population of means of n data (σ_n) is related to the population standard deviation (σ) by

$$\sigma_n = \frac{\sigma}{\sqrt{n}} \tag{2.4}$$

In the example given $n=4$, therefore $\sigma_4=\sigma/2$. Just as s estimates σ, $s/\sqrt{n}$ estimates $\sigma/\sqrt{n}$. The so-called sample standard deviation of the mean is therefore defined by

$$s_n = \frac{s}{\sqrt{n}} \tag{2.5}$$

The standard deviation of the mean thus also becomes smaller as the number of data increases, reflecting our increasing confidence in the value of the mean (see figure 2.4).

This answers the second question posed above (are more repeats better than fewer repeats?). By taking more and more data the mean becomes a better estimate of the population mean because its standard deviation is smaller as the number of data increases. In fact as n

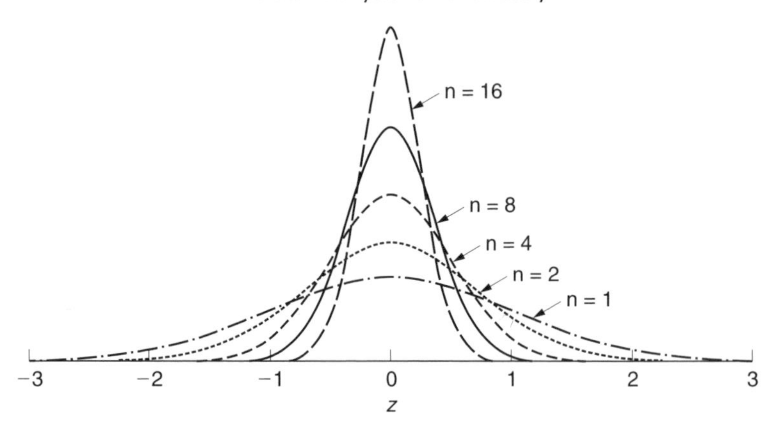

Figure 2.4 Normal probability distribution functions for means of 1, 2, 4, 8, and 16 data.

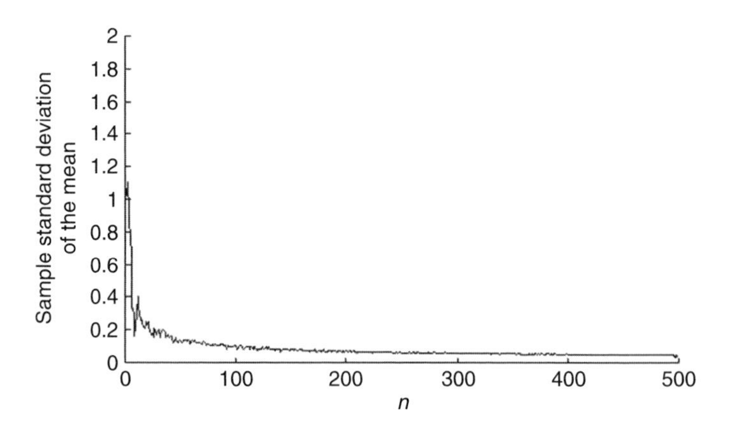

Figure 2.5 Sample standard deviations of the mean of n data randomly drawn from a normally distributed population with $\mu = 10$ and $\sigma = 1$.

approaches infinity, the sample standard deviation of the mean goes to zero. Figure 2.5 shows this happening. The square root of n in equations 2.4 and 2.5 deals a small blow. If it is desired to halve the sample standard deviation of the mean, four times the number of experiments have to be done. Improvement by a factor of 10 implies 100 times more experiments. Therefore, how many repeats are performed depends on what the result will be used for. Once again the idea of fit for purpose comes into play; the precision is dictated by how many experiments you are willing to do.

The central limit theorem also delivers another positive for the analyst. Most of the simple data analysis assumes a normal distribution of data. Much of the time for real sets of data this is not so, but by taking averages of results the distribution of the means tends to a normal distribution, even if the original population is not normally distributed. Hence taking averages of data also helps us with data analysis by removing concerns we might have had about whether our data conform to a normal distribution.

2.5 So How Do I Quote My Uncertainty?

2.5.1 Confidence intervals and confidence limits

The standard deviation of the mean tells all there is to know about the dispersion of the data and it is sufficient to quote the mean and sample standard deviation of the mean (and the number of data). However, it may be more immediately informative to give a range of values that would encompass some proportion of repeated data (say 95 or 99%). Remember that the normal distribution goes from plus infinity to minus infinity but there is not much to be gained by saying that you are 100% sure the right answer is somewhere! What we do, therefore, is to accept some doubt, and give a finite range in which a large proportion of repeated intervals are expected to contain the true value. It is quite easy to give such a defined range (called a confidence interval) because we know exactly what proportions of the normal distribution fall at given multiples of the standard deviation about the mean (see section 1.8.2).

2.5.2 When you have a lot of data: Confidence interval knowing the population mean and standard deviation

For a normal distribution with population mean μ and standard deviation σ the symmetric interval about the mean containing a fraction $1-\alpha$ of the results is given by

$$\mu \pm z_{\alpha/2}\sigma \tag{2.6}$$

where $z_{\alpha/2}$ is obtained from tables of the normal distribution. In other words $z_{\alpha/2}$ is the number of standard deviations either side of the mean containing a fraction $1-\alpha$ of the distribution. For example, for

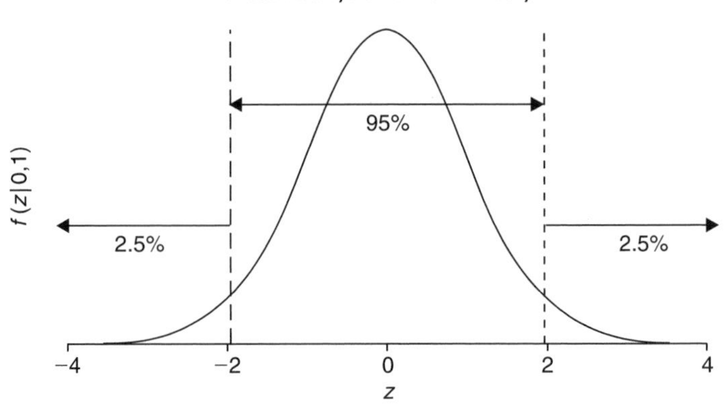

Figure 2.6 Probability distribution function of the normal distribution with dashed lines drawn at $\pm 1.96\sigma$. The area between the lines contains 95% of the distribution.

a 95% confidence interval $\alpha = 0.05$. Another interpretation is to say the probability of finding a value outside the 95% confidence interval is 0.05. The subscript $\alpha/2$ is written because the interval covering, for example, 95% about the mean has $\alpha = 0.05$ with 0.025% at one end and 0.025% at the other (see figure 2.6).

When we are dealing with means of n repeat measurements, the interval containing $(1 - \alpha)$ of the means is given by the standard deviation of the mean, σ_n (see equation 2.4):

$$\mu \pm z_{\alpha/2}\sigma_n = \mu \pm \frac{z_{\alpha/2}\sigma}{\sqrt{n}} \tag{2.7}$$

Comparing equations 2.6 and 2.7 we see that performing n repeated experiments reduces the interval by $1/\sqrt{n}$. In the case of the 95% confidence interval, $\alpha/2 = 0.025$, $z_{0.025} = 1.96$, and thus

$$\mu \pm \frac{z_{0.025}\sigma}{\sqrt{n}} = \mu \pm \frac{1.96\sigma}{\sqrt{n}} \tag{2.8}$$

Although the interval in equation 2.8 is often expressed as a probability interval for the population mean ("The true value lies in this interval with a probability of 95%"), it is not. The population mean μ is what it is with 100% probability. (The only problem is that we do not know it!) What the confidence interval tells us is that if a large number of repeats of our n experiments were performed under identical conditions 95% of the confidence intervals calculated by equation 2.8 would indeed include the value of μ. In reality it is highly unlikely that having done n experiments to give one mean and confidence interval, you will now repeat the whole process again and again to prove this. The $100(1-\alpha)\%$ confidence interval on the experimental mean $\overline{x}$ is given by

$$\mu - \frac{z_{\alpha/2}\sigma}{\sqrt{n}} < \overline{x} < \mu + \frac{z_{\alpha/2}\sigma}{\sqrt{n}} \qquad (2.9)$$

We can rearrange this equation to give a confidence interval on μ given $\overline{x}$:

$$\overline{x} - \frac{z_{\alpha/2}\sigma}{\sqrt{n}} < \mu < \overline{x} + \frac{z_{\alpha/2}\sigma}{\sqrt{n}} \qquad (2.10)$$

Again the only correct statement about the confidence interval is the one given above: that 95% of repeated confidence intervals will include μ. However, it is common to express this incorrectly as "that with $100(1-\alpha)\%$ confidence, the true mean is between the values defined by the confidence limits."

The confidence interval is given by equation 2.10. The confidence limits are the values defining the interval: $\overline{x} \pm z_{\alpha/2}\sigma/\sqrt{n}$. Because the normal distribution is symmetrical the same value $(z_{\alpha/2}\sigma/\sqrt{n})$ is added to the mean to give the upper confidence limit and is subtracted from the mean to give the lower confidence limit. This may not be the case for other distributions; for example, the lognormal distribution has a longer tail on the high side of the mean. See section 2.5.5 to learn under what circumstances this calculation is applicable.

2.5.3 When you only have a small amount of data: Confidence interval knowing the sample mean and standard deviation

There is a problem with equation 2.10 in that although we know $\overline{x}$, n, and $z_{\alpha/2}$, if we only have our n data, and n is small, we do not know σ. We can calculate the sample standard deviation s, but just as $\overline{x}$ is an estimate of μ, s is only an estimate of σ. W.S. Gossett, while working for Arthur Guinness's brewery in Dublin, Ireland, in 1908, published a paper using the pseudonym "Student," which solved this problem. The t-value or Student t-value is used to determine the confidence interval for samples of finite size for which only the sample standard deviation is known. Thus in equation 2.10, $z_{\alpha/2}\sigma$ is replaced by $t_{\alpha,\ n-1}s$. The t-value depends on the probability level required (α) and also the degrees of freedom, that is, the number of repeat measurements (a table of t-values is given in the Appendix). A confidence interval of a mean based on the sample standard deviation is therefore

$$\bar{x} - \frac{t_{\alpha,n-1}s}{\sqrt{n}} < \mu < \bar{x} + \frac{t_{\alpha,n-1}s}{\sqrt{n}} \tag{2.11}$$

To illustrate the difference between these two calculations, look at figure 2.7, which shows the difference between the 95% point on the normal distribution ($z = 1.96$) and the corresponding Student t-value for different degrees of freedom. The t-value depends on the degrees of freedom, and asymptotically approaches the value of z as the degrees of freedom tends to infinity. The important point is to note how quickly this happens. For a sample of three measurements there are two degrees of freedom and $t_{0.05,2}$ is 4.3, more than twice $z_{0.025}$. Therefore a confidence interval based on a standard deviation of three results will be twice that if the population standard deviation, σ is

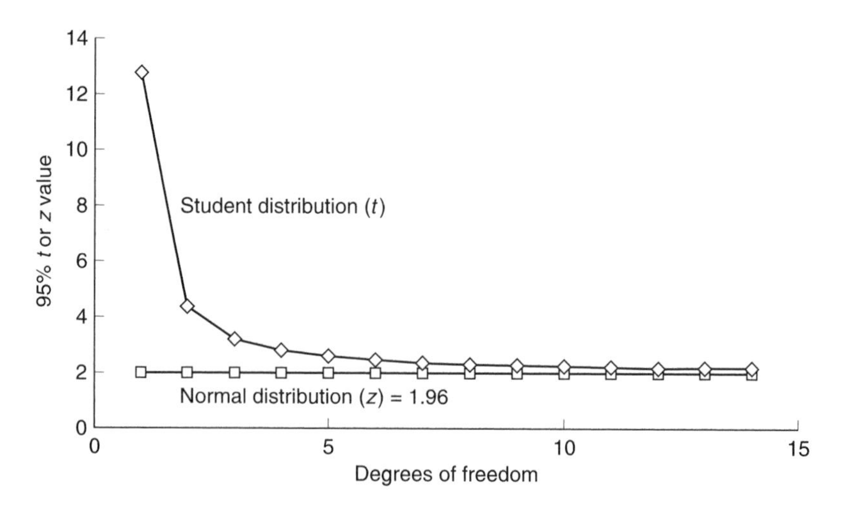

Figure 2.7 Plot of the Student t-value for calculation of a 95% confidence interval with increasing degrees of freedom. The corresponding z-value from the normal distribution is shown ($z_{0.025} = 1.96$).

known. However, as more results are taken the difference between t and z becomes less. By the time we have 10 degrees of freedom the extra width of the confidence interval is only 14%.

Figure 2.8 shows the confidence intervals calculated for the means of the random data used earlier. Figure 2.8(a) shows 95% confidence intervals based on the population standard deviation (which we know: $\sigma = 1$) and z-value (1.96), which is of course the same for each value. Figure 2.8(b) is the 95% confidence interval calculated using equation 2.11. For small values of n the Student t interval is much greater than the one based on a knowledge of σ, because, as discussed above,

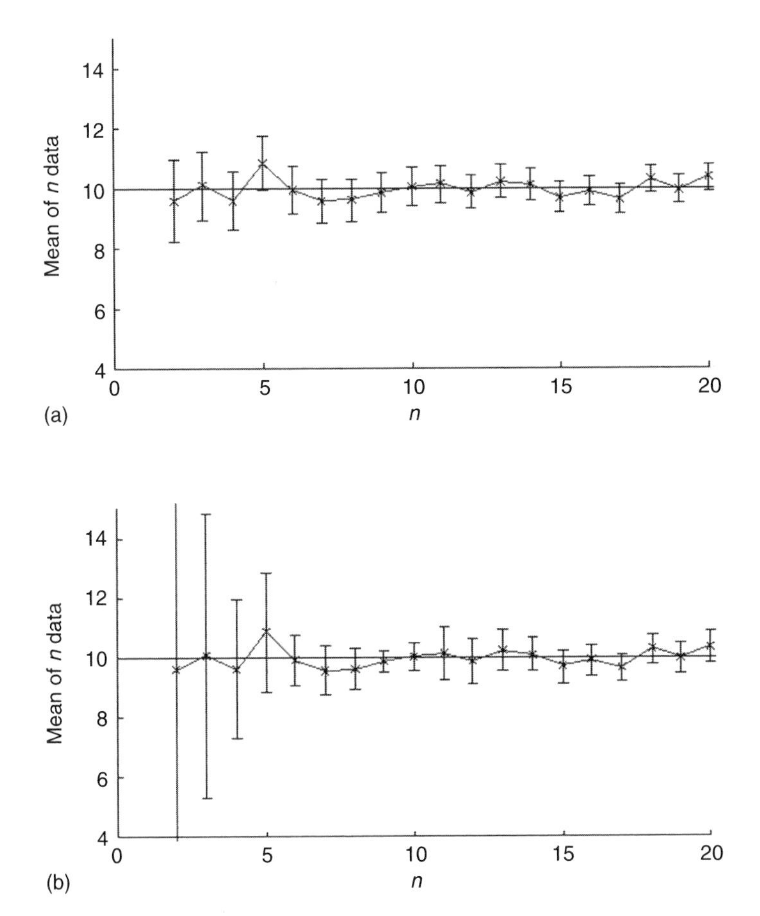

Figure 2.8 95% confidence intervals on the means shown in figure 2.1: (a) calculated from the population standard deviation and z-value (equation 2.10); (b) calculated from the sample standard deviation and Student t-value (equation 2.11).

s calculated from n data is only an imperfect estimate of σ. In fact for one degree of freedom the interval is off the scale of the graph at ±10. The moral of this tale is do not bother calculating standard deviations and confidence intervals if you only have two or three data; some say you should have at least five to eight.

2.5.4 Tails

An unnecessary complication, which was possibly once introduced to make life easier, is the distinction between one- and two-tailed Student t-values (tails are also used in other statistics). Two-tailed probabilities are spread over the two ends of the distribution with half the given probability in each tail, and are denoted by putting a double prime ($''$) after the probability value. One-tailed probabilities are shown as a single prime ($'$) and refer to just one tail of the distribution. For example, for a 95% confidence interval and 10 degrees of freedom, $t_{0.025',10}$ is equal to $t_{0.05'',10}$, as can be seen from figure 2.9. Annoyingly, in Excel the z values obtained from the normal distribution are always one tailed (=−NORMSINV(p)[1]) but the Student t-values

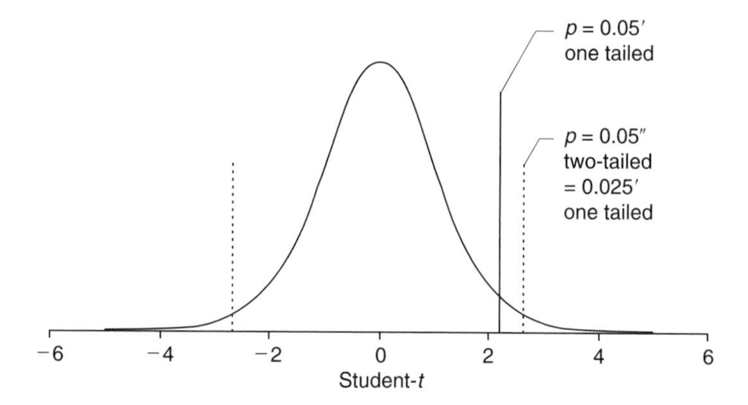

Figure 2.9 Student t-distribution (T) for 10 degrees of freedom. Solid vertical line: value for which $p(T > t) = 0.05$, that is, $t_{0.05',10}$ (one-tailed). Dashed lines: values for which $p(T > t, T < -t) = 0.05$, that is, $t_{0.05'',10}$ (two-tailed) $= t_{0.025',10}$ (one-tailed).

[1]NORMSINV(p) returns the x-value at which the area under the normal distribution pdf (with $\mu = 0$ and $\sigma = 1$) from $-\infty$ to x is p. Therefore a negative x is returned as the area calculated is at the left-hand tail of the distribution.

(=TINV(*p*,*df*)) are two tailed. However, when the probability of finding a Student t (T) greater than a particular t is to be calculated using TDIST, whether the one-tailed or two-tailed values are to be returned must be specified (=TDIST(*t*, *df*, *tails*)).

Example 2.2

Determine a probability ($T > t$) from a t-value using Excel.

Problem

What is the probability associated with a two-tailed Student t-value of 2.23 with 10 degrees of freedom?

Solution

This can simply be done in Excel using the function TDIST with syntax TDIST(*t*, *df*, *tails*). Therefore in a blank cell of a spreadsheet you would input =TDIST(2.23, 10, 2).

Answer

The probability associated with a two-tailed Student t-value of 2.23 with 10 degrees of freedom is 0.0498.

Comments

1. If =TDIST(2.23, 10, 1) were typed into a cell in the spreadsheet then the output would be 0.024921 (i.e., 0.025) as this refers to a one-tail probability.
2. The probability 0.050 is the value of α used in calculating the percentage of data falling within a confidence interval of $100(1-\alpha)\%$. Hence 2.23 is a t-value for a 95% confidence interval.
3. The answer is given to three significant figures following the number of significant figures in the question (2.23 has three significant figures).

2.5.5 After how many measurements can you assume $s = \sigma$?

It would be a lot easier if we could ignore Student t-values and just assume that the s calculated from our data is σ. The value of z from the normal distribution for the 95% confidence interval is, as we have learned, 1.96. For 30 degrees of freedom, the error in using z and not t ($t_{0.05'',30} = 2.04$) is about 4%. Hence the answer is that it depends on the error you can tolerate, but you should usually consider the Student t-distribution for less than about 30 data.

2.5.6 How do I write standard deviations and confidence intervals?

Whenever you write a result and decide to include a measure of precision (as you always should) it is important to convey enough information for the user of the result to assess the precision. Plus and minus (±) should be reserved for a confidence interval, because an interval is what ± defines. The probability should be given somewhere. It is not sufficient to assume the reader will know it is 95%; many authors put ± before one sample standard deviation of the mean, which encompasses only 68% of the distribution. The degrees of freedom are allowed for in the value of t but it is good practice to mention n. If you want to quote a standard deviation, it should be given in parentheses and must also show the value of n. See the answer for example 2.1b below for the appropriate way of presenting an uncertainty with a confidence interval.

Example 2.1b

Problem

Calculate the mean and 95% confidence limits for the data in example 2.1a.

Solution

In Example 2.1a the injection precision is determined by calculating the mean, standard deviation, and RSD. We can now use these values to obtain confidence limits at the required probability. In this example the 95% confidence limit for the injection precision data in example 2.1a is calculated.

Calculating by hand

Recall the confidence interval is given by $\bar{x} \pm t_{\alpha,n-1}s/\sqrt{n}$ with, in this case, $\alpha = 0.05''$, $n = 6$, $\bar{x} = 2919564$, and $s = 525705$.

From the table of t-values (see Appendix) for five degrees of freedom (as one degree of freedom is used up in the calculation of the mean) the value of $t_{0.05'',5} = 2.57$. Therefore

$$\bar{x} \pm \frac{t_{0.05'',5}s}{\sqrt{n}} = 2919564 \pm \frac{2.57 \times 525705}{\sqrt{6}} = 2919564 \pm 551569$$

Answer

The mean and 95% confidence interval of the peak areas are $(2.92 \pm 0.55) \times 10^6$ ($n = 6$).

Calculation using Excel

To calculate the 95% confidence limits using Excel, simply type into a blank cell =STDEV(*range*)*TINV(0.05, *n*−1)/SQRT(*n*), where *range* is the range of cells containing the *n* data. Typing =TINV(0.05, *n*−1) gives the t-value alone.

Hence for the set of data the spreadsheet will look like spreadsheet 2.2.

Answer

The mean and 95% confidence interval of the peak areas are $(2.92 \pm 0.55) \times 10^6$ ($n = 6$).

Spreadsheet 2.2

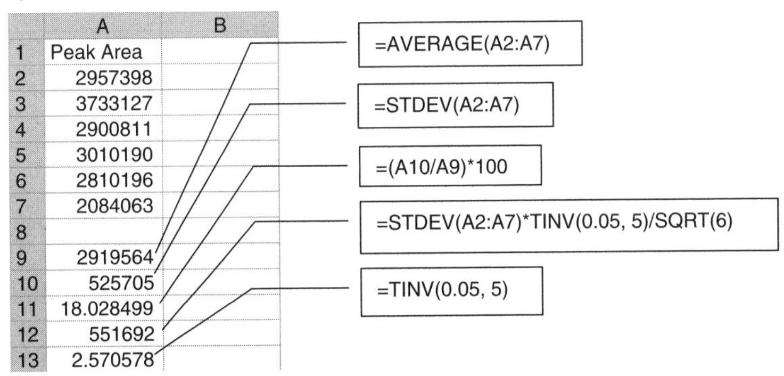

Comments

1. Again, note the wording of the final answer. It was worded this way to inform the reader that it is a 95% confidence limit from 6 samples and hence the t-value that had been used was for 5 degrees of freedom.
2. We recommend writing the standard deviation or 95% confidence interval with two significant figures. The mean is then quoted to the same significance (decimal places).
3. With very large or very small numbers, to avoid many zeros that can be easily misread, use prefixes for units (e.g., μM, nm, MPa) or use scientific notation by multiplying by 10 to a suitable power as we have done here.
4. Note the value of the 95% confidence limit, before rounding, is larger than that obtained when calculating by hand. This is simply because the spreadsheet uses the t-value of 2.57057763519696 while when calculating by hand we used the truncated, tabulated value of 2.57. Despite this small difference the answer is the same once we limit the confidence interval to two significant figures.
5. As Excel only gives a two-tailed result from TINV, we must use $\alpha'' = 0.05$.

2.5.7 How many measurements?

When establishing confidence intervals from data, analysts are sometimes told to make seven measurements. The magical nature of this number stems from the fact that the standard deviation of seven results is just greater than the 95% confidence interval of the mean. How so? Because the 95% confidence interval for $n = 7$ is $t_{0.05'',6}\, s/\sqrt{7} = (2.45\, s)/2.65$ ($t_{0.05'',6} = 2.45$ and $\sqrt{7} = 2.65$), and so these nearly cancel leaving s as about the 95% confidence interval of the mean.

The answer to how many measurements *you* should make is answered by how precise you want your result. If the population standard deviation is known, then if you want to make sure that the error in your mean, that is, the difference between the mean and the population mean that would be obtained by an infinite number of repeats, is no more than ε with a probability of α then

$$\varepsilon = \mu - \overline{x} = z_{\alpha/2}\frac{\sigma}{\sqrt{n}} \tag{2.12}$$

Equation 2.12 may be rearranged to give

$$n = \left(\frac{z_{\alpha/2}\sigma}{\varepsilon}\right)^2 \tag{2.13}$$

Equation 2.13 is instructive. It tells us that we must do more experiments if we want a smaller uncertainty (ε), a higher probability (α), or have a smaller standard deviation of our experiments.

If we do not know the population statistics μ and σ but only have data from some preliminary experiments then we are in something of a dilemma. First, because we have done some experiments perhaps we have already done too many. Second, the equivalent expression to equation 2.13 is

$$n = \left(\frac{t_{\alpha'',n-1}s}{\varepsilon}\right)^2 \tag{2.14}$$

where s is the sample standard deviation. It is possible to cast equation 2.14 in terms of the relative standard deviation and ε as the relative (%) target error:

$$n = \left(\frac{t_{\alpha'',n-1}RSD}{\varepsilon\%}\right)^2 \tag{2.15}$$

The difficulty with using equation 2.14 or 2.15 is that we need to know n to calculate the degrees of freedom, to give the Student t-value. It is possible to iterate equation 2.14 or 2.15 with an initial guess at n to give t which is then put back in n and so on. After the experiments are performed it may be necessary to recalculate n to take into account the new value for the standard deviation. In practice, we are usually only interested in ball-park figures, for example 5% with anywhere between 4 and 6% being acceptable, and hence the process is not too tedious.

Example 2.1c

The RSD of the injection volume for six injections was a poor 18%.

Problem

How many injections are required to reduce the relative error of the mean to 5% with 95% confidence?

Solution

To solve this problem requires using equation 2.14. As the value of $t_{0.05'', n-1}$ used is dependent on the value of n we must make a guess at n to get $t_{0.05'', n-1}$ and then calculate a new value of n. The new n is then input back into equation 2.14 and this iterative process is repeated until a value of n is converged upon. This process could be performed by hand but is far simpler to perform in Excel as shown below.

Calculation using Excel

1. Input the values of the target error, ε, and the calculated RSD (in this case 5 and 18, respectively) into cells as shown in spreadsheet 2.3.
2. Make an initial guess of the number of experiments required. In the spreadsheet our initial guess is the number of experiments performed, that is, 6, and hence the degrees of freedom is 5.
3. Use the number of degrees of freedom and the probability to calculate $t_{0.05'', n-1}$ using =TINV.
4. Define two columns as the number of degrees of freedom, *df*, and the number of experiments n. In the first cells type "5" below *df* and =ROUND((TINV(0.05, A21)*$B$17/$B$16)^2,0) below n. This expression is equation 2.14, where A21 refers to *df* and ^2 squares the function. The rounding function with syntax ROUND(*number*, *decimal_places*) rounds the answer to a prescribed number of decimal places. In this case, because you cannot do half experiments,

the expression is rounded to 0 decimal places giving an integral number of experiments to perform. The spreadsheet looks like spreadsheet 2.3.

Spreadsheet 2.3

	A	B
16	Error	5
17	RSD	18
18	t-value	2.57
19		
20	df	n
21	5	86
22	85	51
23	50	52
24	51	52
25	51	52
26	51	52

=TINV(0.05, $A$21)

=ROUND((TINV(0.05, A21)*$B$17/$B$16)^2,0)

=(B21-1)

=ROUND ((TINV(0.05, A22)*$B$17/$B$16)^2,0)

Answer

The number of experiments required to have 95% confidence that the error in the injection precision is 5% or less is 52.

Comments

1. If this were a real-life problem, we would be in a dilemma. It is unlikely each measurement could be repeated 51 times, so our only recourse would be to renegotiate the target uncertainty upwards (perhaps to 10% when "only" 15 replicates need be done), or to revise our procedure to improve the precision.

2.6 Robust Estimators

The average and sample standard deviation are known as "estimators" of the population mean and standard deviation. We have seen how the estimates improve as the number of data increases. As we have stressed, the use of these statistics requires data that are normally distributed, and for confidence intervals employing the standard deviation of the mean this tends to be so. Real data may be so distributed, but often the distribution will contain data that are seriously flawed, as with the RACI titration competition described in chapter 1. If we can identify such data and remove them from further consideration, then all is well. Sometimes this is possible, but not always (the statistically valid identification of data as outliers is discussed in chapter 3). This is a problem as a single rogue value can seriously upset calculations of the mean and standard deviation. Estimators that can tolerate a certain amount of bad data are called *robust estimators* and can be used when it is not possible to ensure that the data being processed has the correct characteristics. Here we shall introduce the middle value of an ordered set of data (median) as a robust estimator of the mean, and the range of the middle 68% of the data (normalized interquartile range) as a robust estimator of the standard deviation. Robust methods have their place, particularly when we must keep all the data together in, for example, an interlaboratory trial where an outlying result from a laboratory cannot simply be ignored. However, robust estimators are not the best statistics and wherever possible the statistics appropriate to the distribution of the data should be used.

2.6.1 Median

The median is the middle value of a set of data when arranged in ascending order. If there are an odd number of data then there is a unique middle datum. If there are an even number then the median is the average of the middle two data. It is robust, because no matter how outrageous one or more extreme values are they are only individual values at the end of a list. Their magnitude is immaterial. The RACI data shown in table 1.2 are ordered in table 2.1

2.6.2 Interquartile range

The interquartile range (IQR) is the range of values that spans the middle 50% of data. Three quarters of the IQR, known as the

Table 2.1 The results of 25 competitors in the 1997 RACI titration competition of the concentration of a test acetic acid solution in units of mol L^{-1}

0.0920, 0.0936, 0.1134, 0.1138, 0.1139, 0.1141, 0.1142, 0.1143, 0.1143, 0.1144, 0.1144, 0.1145, **0.1146**, 0.1148, 0.1150, 0.1150, 0.1152, 0.1153, 0.1155, 0.1158, *0.1177, 0.1219, 0.1222, 0.1556, 0.9083*

normalized IQR, is an estimate of the standard deviation[2]. A problem with the IQR is that it cannot be calculated for small data sets, as there have to be sufficient data to define quartiles (sections of the ordered data that contain one-quarter of the data).

Example 2.3

Calculation of robust estimators.

Problem

Use robust estimators to estimate the population mean and population standard deviation for the RACI titration competition data shown in table 2.1. Outliers from a normal distribution are shown in italics (see section 3.4 for details of how to do this) and the median is in bold.

Solution

The median is the middle value when the data are sorted into ascending order. In this case there are 25 data and hence the 13th datum is the median which is 0.1146 M.

The interquartile range is the middle 50% of the data. As there are 25 data points we take the middle 13 from 0.1142 to 0.1155 which gives an interquartile range of 0.0013 M. Therefore the normalized interquartile range is $0.0013 \times 0.75 =$ 0.00098 M.

Answer

For the RACI titration competition data the median is 0.1146 M and the normalized interquartile range is 0.00098 M.

[2] The IQR encompasses 50% of the data; a spread of two standard deviations about the mean (± 1) includes 68% of the data which leads to $\sigma = 0.73 \times \text{IQR}$, which for ease of calculation is rounded to $\frac{3}{4} \times \text{IQR}$.

Comments

1. The assigned value in the titration competition was 0.1147 M, which makes the median (0.1146 M) a good estimate of the mean, even with seven identifiable outliers. The average, on the other hand, is 0.1470 M which is hopelessly skewed to higher values by the last value of 0.9083 M. When the outliers are not used, the mean is also 0.1146 M.
2. The sample standard deviation of the whole RACI data set is 0.1590 M, again greatly inflated by the high outliers. The standard deviation suffers more than the mean from outliers because of the squaring of the difference between the value and the mean (equation 2.2). The normalized IQR of the whole data set is 0.00098 M and of the truncated set (minus the seven outliers) is 0.00058 M. The standard deviation of the truncated set is 0.00063 M.

2.7 Repeatability and Reproducibility of Measurements

What happens in real analytical laboratories? There is very rarely sufficient time or resources to perform experiments enough times on each test material to establish a reasonable mean and standard deviation (and therefore confidence interval). Duplicate measurements are the norm rather than the exception.

Many working laboratories will have performed a similar analysis many times over, whether it be the analysis of active ingredients for a pharmaceutical production line, or the analysis of an element in an ore for a mining company. When the analytical method was first established, method validation will have determined repeatability and reproducibility standard deviations, and these will have been verified for use in the particular laboratory.

The repeatability standard deviation is defined as "The precision of a method expressed as the standard deviation of independent determinations performed by a single analyst using the same apparatus and techniques." Hence the repeatability is what governs any replicate measurements made in your laboratory by you on the test material. The reference to "independent measurements" means that a number of separate test portions should be weighed, dissolved, and measured, not that the same solution should be presented to the analytical instrument a number of times. (If this were to be done the standard deviation of these measurement results would be the repeatability of the instrument measurement, not the repeatability of the analysis.)

Reproducibility standard deviation is "The precision of a method expressed as the standard deviation of determinations performed in different laboratories." Remembering the discussion of chapter 1 we would expect the reproducibility to be greater than the repeatability, as each laboratory will have its own repeatability (which might be expected to be about the same) but the differences between the laboratories reflecting different biases will now add to this to give the reproducibility. Experience has shown that the interlaboratory reproducibility is about two to three times the repeatability.

For completeness it should be noted that "intralaboratory reproducibility" is sometimes used to refer to the standard deviation of measurement results obtained within the same laboratory, but perhaps by different analysts and/or different equipment and/or different days.

Returning to the practical problem of routine measurement, if a laboratory has established a reasonable estimate of the repeatability standard deviation, then the routine duplicate measurements may be checked against this value. Suppose the repeatability standard deviation is σ_r. The standard deviation of the difference between two measurements is $\sqrt{2}\sigma_r$ and the 95% confidence interval on the expected difference of 0 is $\pm 1.96 \times \sqrt{2}\sigma_r = 2.8\sigma_r$.

Therefore if the difference between duplicated results measured under repeatability conditions is greater than $2.8\sigma_r$ there should be concern that there is something wrong with the analysis. This can be used as part of a quality control procedure to ensure consistency of results. An equivalent difference can be defined for a reproducibility standard deviation (σ_R) for checking results found between laboratories. The maximum permissible difference $2.8 \times \sigma_r$ is known as the *repeatability limit* (r) and $2.8 \times \sigma_R$ is the *reproducibility limit* (R).

3

Hypothesis Testing

3.1 What This Chapter Should Teach You

- To understand the concept of the null hypothesis and the role of Type I and Type II errors.
- To test that data are normally distributed and whether a datum is an outlier.
- To determine whether there is systematic error in the mean of measurement results.
- To perform tests to compare the means of two sets of data.

3.2 Why Perform Hypothesis Tests?

One of the uses to which data analysis is put is to answer questions about the data, or about the system that the data describes. In the former category are "is the data normally distributed?" and "are there any outliers in the data?" (see the discussions in chapter 1). Questions about the system might be "is the level of alcohol in the suspect's blood greater than 0.05 g/100 mL?" or "does the new sensor give the same results as the traditional method?" In answering these questions we determine the probability of finding the data given the truth of a stated hypothesis—hence "hypothesis testing."

A hypothesis is a statement that might, or might not, be true. Usually the hypothesis is set up in such a way that it is possible to calculate the probability (P) of the data (or the test statistic calculated from the data) given the hypothesis, and then to make a decision about whether the hypothesis is to be accepted (high P) or rejected (low P). A particular case of a hypothesis test is one that determines whether or not the difference between two values is significant—a significance test. For this case we actually put forward the hypothesis that there is no real difference and the observed difference arises from random effects: it is called the null hypothesis (H_0). If the probability that the data are consistent with the null hypothesis falls below a predetermined low value (say 0.05 or 0.01), then the hypothesis is rejected at that probability. Therefore, $p < 0.05$ means that if the null hypothesis were true we would find the observed data (or more accurately the value of the statistic, or greater, calculated from the data) in less than 5% of repeated experiments. To use this in significance testing, a decision about the value of the probability below which the null hypothesis is rejected, and a significant difference concluded, must be made. So what is the predetermined low value of the probability at which we decide to reject H_0 and how do we calculate the actual probability of finding the data given H_0? If you remember confidence limits—the range of values in which a certain percentage of results should fall—we should be able to use them to decide if the results we are comparing are near enough together. The null hypothesis is rejected "at the 95% level of confidence" if the probability of the test statistic, given the truth of H_0, falls below 0.05. In other words, if H_0 is indeed correct, less than 5% (i.e., 1 in 20) means of repeated experiments would fall outside the limits. In this case it is concluded that there was a significant difference. If we think about a normal distribution with a 95% confidence interval, it is so called because 95% of repeated measurement results would fall inside the interval and 5% outside. Therefore if we were to reject H_0 at the 95% probability level we are admitting that 5% (i.e., 1 in 20) of repeated experiments would be rejected in error when they were really part of the distribution.

3.3 Levels of Confidence and Significance

A criminal court of law needs to be convinced of a defendant's guilt "beyond all reasonable doubt." Civil cases are decided "on the balance of probabilities." What do these statements mean? Courts are very careful to avoid putting actual figures on these statements, but the first (beyond all reasonable doubt) might be, say, 99.9% certain while the second might only be 60% or less. If we have the luxury of having data that do allow real probabilities to be determined then it is possible to give the court the chance to decide what is "reasonable doubt." Of course, criminal cases are extremely complex and although, for example, the probability of a DNA match may be given in a proper statistical fashion, the myriad of other evidence will still need to be weighed by the jury without such help.

Many tests that are published in the chemical literature compare a calculated value of a statistic with tables of critical values for this statistic at a given probability (often $p = 0.05$, i.e., at the 95% level). Papers in the literature often include statements such as "the results were not significantly different (95% level)," or "there was a significant difference with 95% probability," and although there are better ways of expressing the statistically correct statement, we do receive the message about the relationship between the sets of data. The figure of 95% is a somewhat arbitrary one, arising because of the accident that $\mu \pm 2\sigma$ covers about 95% of a population. With modern computers and spreadsheets it is possible to calculate the probability of the statistic given a hypothesis, leaving the reader to decide whether to accept or reject it. This approach is recommended in this text, and hopefully will lead to a more considered view of hypothesis testing.

In deciding what is a reasonable level to accept or reject a hypothesis, that is, how significant is "significant," two scenarios, in which the wrong conclusion is arrived at, need to be considered. First, is the case in which we reject a hypothesis when it is actually true (a so-called Type I error). In biosciences a Type I error is often referred to as a false negative. Here, the conclusion of the significance test is that the difference being tested is outside a reasonable range of what would be expected of a normal distribution consistent with the null hypothesis when in fact H_0 is true. The second scenario is the opposite of this, when the significance test leads to the analyst wrongly accepting the null hypothesis although in reality H_0 is false (a Type II error). Type II errors will be familiar to bioscientists as false positives. These are discussed further below.

Consider what the consequences of setting the probability level for acceptance of H_0 at 90, 95, and 99% might be. As an example suppose an analytical method has been used to analyze a certified reference material for the element zinc, that is, a material whose amount of substance of zinc has been established to a high metrological standard with low measurement uncertainty, with a view to deciding if there is any significant systematic error in the method. The mean of n measurement results has been determined and suppose that the population standard deviation (σ), and therefore the standard deviation of the

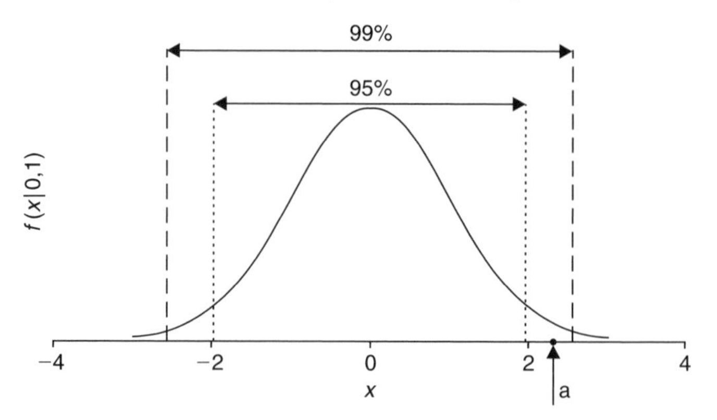

Figure 3.1 Normal distribution showing a result (a) that falls between the 95% and 99% limits. In the example given in the text, $x = 0$ corresponds to the value of the certified reference material and the axis shows values of standard deviations of the mean from this value.

mean ($\sigma/\sqrt{n}$), is known. The normal curve in figure 3.1 is the pdf of the population of means (see chapter 1 for a description of the normal pdf) and the vertical lines indicate between which values of x different percentages of the population lie. The x-axis shows the standard deviation of the mean. Consider a result at the value labeled a. It is between the 95% limit and the 99% limit. Deciding that there is a significant systematic error, and so to reject the null hypothesis "that there is no significant difference between the amount of substance of zinc in the certified reference material determined by the analytical method and the amount of zinc actually certified" at a particular probability requires the measured mean to be outside the range defined by the limits. In the example we would conclude that there was a significant systematic error at the 95% level, but not at the 99% level. It is possible to calculate the probability at which the mean is significantly different as 98%. We know therefore that only 2% (100–98) of means determined by a method without systematic error would be as far or farther away from the certified value by an amount $a\sigma/\sqrt{n}$, with 1% greater and 1% less than the range. By choosing the limits to include $100 \times (1 - \alpha)\%$ of the distribution, the probability of making a Type I error introduced above is $100 \times \alpha\%$, here 2%.

It seems that to make a small as possible Type I error (that is, wrongly rejecting the null hypothesis) all we have to do is to set higher and higher probability limits by choosing smaller and smaller α. This is up to the analyst, but as the limit for rejecting H_0 is pushed farther

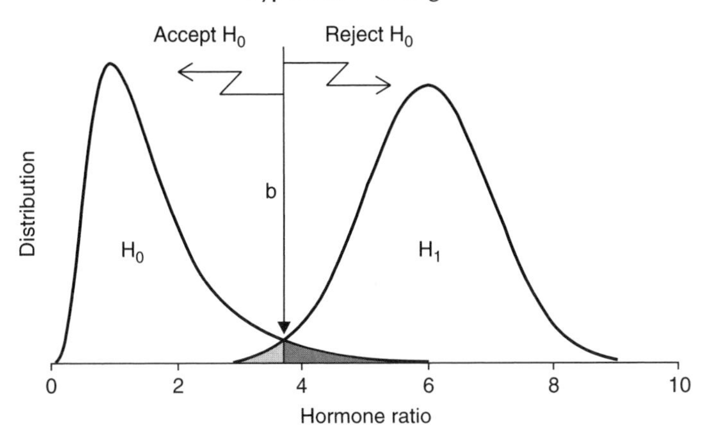

Figure 3.2 Distribution of ratios of hormones in the general public (left) and in a group of athletes who have taken it as a performance-enhancing drug (right). Point **b** is a ratio at which the probability of making a Type I or Type II error is minimized.

out (90, 95, 99, 99.9%...) we run an increasing risk of the other error—the Type II error of accepting H_0 when it is false. Making a Type II error arises when the result we are testing is determined to belong to the population defining the null hypothesis, when in fact belongs to some other population. Here lies a problem as our null hypothesis population is always defined and α is always set, but rarely do we know much about any of the other alternative populations (with hypotheses H_1, H_2, etc.). Hence usual tests allow us to accept or reject H_0 but say nothing about alternatives.

As an example, observe the distributions of a hypothetical hormone in the blood of healthy male athletes and in that of males given hormone patches during a study (figure 3.2). Obviously the group that has taken the drug has mostly higher levels, but there is some overlap of the distributions. A person found to have an amount labeled b in the figure may be an unusually naturally hormonal athlete or a drug cheat who has not taken very much. Values to the left of b favor the normal population and those to the right favor the drug-taking population. The shaded area to the right of b gives the probability of making a Type I error (α) and the shaded area to the left of b gives the probability of making a Type II error (β) (remember that the area under a pdf is proportional to the probability). Where should we draw the line when testing athletes? The null hypothesis, H_0, is that a result comes from the population who do not take drugs. In terms of this example a Type I error leads to the conviction of an innocent athlete, with all the attendant publicity, loss of livelihood, and public disgrace. A Type II error allows a drug cheat to get away with it. Ideally we choose our method of chemical analysis so that the two populations are far apart and so a measurement result clearly falls within one population or the other. If this is not possible, many sporting authorities accept that the innocent athletes must be protected and require odds of around 1:30,000 against before the null hypothesis is rejected and the athlete is prosecuted, implying a value for α of 3.3×10^{-5} or certainty at the probability level of 99.997% (nearly 4σ) before rejecting H_0 (= the athlete is a member of the normal, nondrug-taking population). If there is overlap between the normal and alternative (drug taking) populations, then there will be a number of drug cheats who will get away with it, the price paid for protecting the innocent.

The point about this example is that all analytical chemistry should be fit for purpose. When you make a decision based on a statistical test, the choice of the probability level at which the null hypothesis is rejected is made by the user, not by a book or software package. Do not adopt probability levels blindly, but consider the risk of making the different types of error.

3.4 How to Test If Your Data Are Normally Distributed

Many of the statistics used by analytical chemists are based on the assumption that the data are normally distributed. Sometimes, in the case of the standard deviation of a mean, the data tend to the normal distribution by theory, but most of the time we cross our fingers and use the normal statistics anyway. Although there is no method that can take three or four values and make a sensible statement about their distribution, there are tests for the assumption of normality for sets of data of at least 10 tests.

A useful graphical procedure to test the normality of a set of data that can be implemented in a spreadsheet is the Rankit method. The procedure, shown in example 3.1 below, is as follows for n data:

1. Sort data into increasing order of magnitude.
2. Write the cumulative frequency of each value, that is, how many data have an equal or lesser value.

3. Calculate the normalized cumulative frequency = cumulative frequency/$(n+1)$.
4. Calculate the value of the normal pdf associated with the normalized cumulative frequency of each value ($=z$).
5. Plot z against the value.

In days gone by this was achieved using "probability paper," specially ruled graph paper which took care of the normal pdf. Nowadays, spreadsheets have functions to perform this calculation: in Excel it is NORMSINV(*x*), where x is the normalized cumulative frequency. If the data are normally distributed this graph should be linear. Obvious outliers are seen as points at the extremes of the x-axis, that is, at values much greater than would be expected. Example 3.1 shows how to determine whether data are normally distributed using a Rankit plot in Excel.

Example 3.1

What results out of the RACI titration data shown in table 3.1 are normally distributed?

Solution

For the data in table 3.1, the five steps listed above are performed as shown below (spreadsheet 3.1):

1. The data are sorted in ascending order in column A. If the data were added to the spreadsheet in random order this could be done using the Sort... option in the Data menu. The function RANK we shall use does not need sorted data, but it is useful for us to see the sorted data.
2. The cumulative frequency for each datum is calculated in column B. The cumulative frequency may be calculated using

Table 3.1 The results of 25 competitors in the 1997 RACI titration competition of the concentration of a test acetic acid solution in units of mol L^{-1}. The value in bold is the median

0.0920, 0.0936, 0.1134, 0.1138, 0.1139, 0.1141, 0.1142, 0.1143, 0.1143, 0.1144, 0.1144, 0.1145, **0.1146**, 0.1148, 0.1150, 0.1150, 0.1152, 0.1153, 0.1155, 0.1158, *0.1177, 0.1219, 0.1222, 0.1556, 0.9083*

Spreadsheet 3.1

=COUNT($A$2:$A$26) + 1 – RANK(A2,$A$2:$A$26)

=B2/(COUNT($A$2:$A$26) + 1)

=NORMSINV(C2)

	A	B	C	D
1	Data	Cumulative Frequency	Normalized	Z
2	0.092	1	0.038462	-1.76882
3	0.0936	2	0.076923	-1.42608
4	0.1134	3	0.115385	-1.19838
5	0.1138	4	0.153846	-1.02008
6	0.1139	5	0.192308	-0.86942
7	0.1141	7	0.230769	-0.73632
8	0.1142	7	0.269231	-0.61514
9	0.1143	9	0.346154	-0.39573
10	0.1143	9	0.346154	-0.39573
11	0.1144	10	0.423077	-0.19403
12	0.1144	11	0.423077	-0.19403
13	0.1145	12	0.461538	-0.09656
14	0.1146	14	0.5	5.47E-10
15	0.1148	14	0.538462	0.096558
16	0.115	15	0.615385	0.293381
17	0.115	16	0.615385	0.293381
18	0.1152	17	0.653846	0.395725
19	0.1153	18	0.692308	0.502402
20	0.1155	19	0.730769	0.615141
21	0.1158	20	0.769231	0.736316
22	0.1177	21	0.807692	0.869424
23	0.1219	22	0.846154	1.020076
24	0.1222	23	0.884615	1.19838
25	0.1556	24	0.923077	1.426077
26	0.9083	25	0.961538	1.768824

the RANK function. RANK(*number*, *range*, *order*) returns the rank of *number* within a given *range*. If the parameter *order* is zero or omitted, the rank is as if the range were sorted in descending order. In order to have the correct frequency in the event of a tie it is necessary to use the following formula: =COUNT(*range*) + 1 – RANK (*number*, *range*).

3. The normalized cumulative frequency is given by the cumulative frequency/$(n+1)$. Therefore the normalized cumulative frequency is calculated in Excel using =*cell1*/(COUNT (*range*) + 1), where *cell1* is the cell containing the cumulative frequency. Here, as there are 25 data points, $n+1=26$.
4. z is =NORMSINV(*cell2*) where *cell2* is the cell containing the normalized cumulative frequency. Therefore z is given by =NORMSINV(*cell1*/(COUNT(*range*) + 1)).
5. Plot z-score against the data on a scatter graph.

The Rankit plot is shown in figure 3.3(a), where z is plotted against the determined concentration as an XY (scatter) chart.

Clearly the point at 0.9083 M is not part of the normal distribution as it lies such a long way from a straight line which is defined by the other points. One might suspect other points are

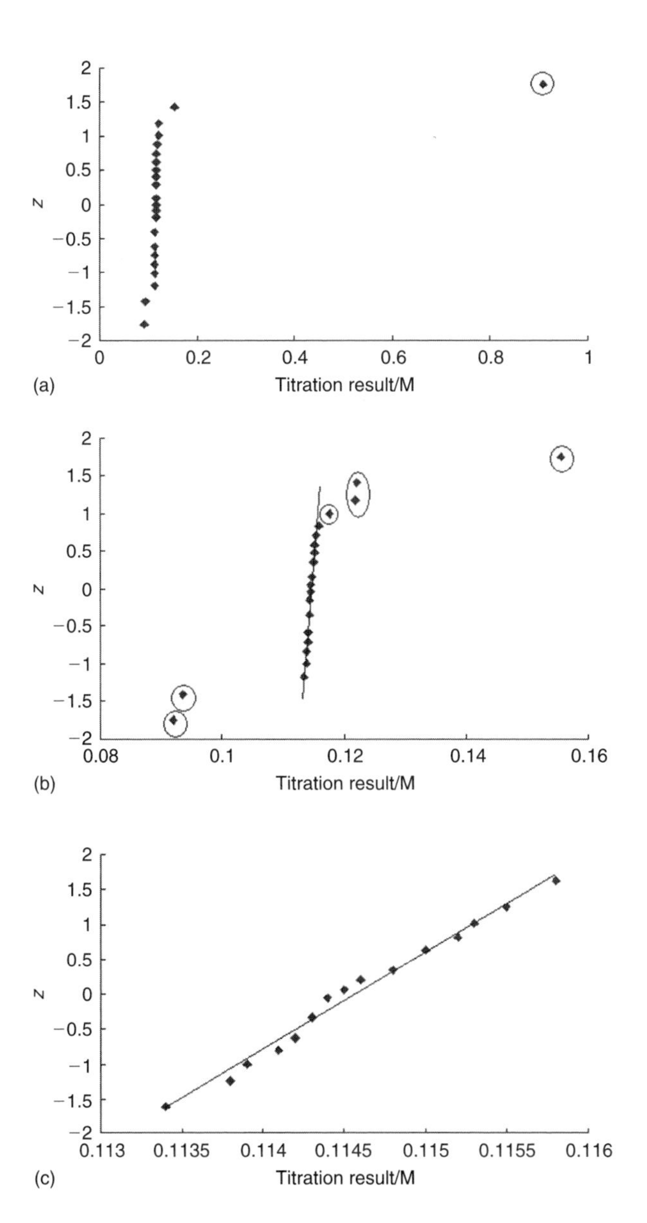

Figure 3.3 Rankit plots for results of the RACI titration competition: (a) All data; (b) with extreme outlier at 0.9083 M removed; (c) with seven outliers removed. Note the shrinking x-axis range.

Spreadsheet 3.2

	A	B	C	D
1	Data	Cumulative Frequency	Normalized	Z
2	0.092	1	0.04	-1.75069
3	0.0936	2	0.08	-1.40507
4	0.1134	3	0.12	-1.17499
5	0.1138	4	0.16	-0.99446
6	0.1139	5	0.2	-0.84162
7	0.1141	7	0.24	-0.7063
8	0.1142	7	0.28	-0.58284
9	0.1143	9	0.36	-0.35846
10	0.1143	9	0.36	-0.35846
11	0.1144	10	0.44	-0.15097
12	0.1144	11	0.44	-0.15097
13	0.1145	12	0.48	-0.05015
14	0.1146	14	0.52	0.050153
15	0.1148	14	0.56	0.150969
16	0.115	15	0.64	0.358459
17	0.115	16	0.64	0.358459
18	0.1152	17	0.68	0.467699
19	0.1153	18	0.72	0.582841
20	0.1155	19	0.76	0.706302
21	0.1158	20	0.8	0.841621
22	0.1177	21	0.84	0.994458
23	0.1219	22	0.88	1.174987
24	0.1222	23	0.92	1.405072
25	0.1556	24	0.96	1.750686

not part of the distribution, so after removal of 0.9083 M as an outlier the process of drawing a Rankit plot is repeated to give spreadsheet 3.2.

The resultant Rankit plot of z plotted against the determined concentration is shown in figure 3.3(b), where again it is apparent that the data points 0.092, 0.0936, 0.1177, 0.1219, 0.1222, and 0.1556 M are outliers.

Removal of these points and again performing the calculations to produce the Rankit plot gives figure 3.3(c), where finally it is clear all points fall on a straight line.

Answer

The 18 points in table 3.1 that are not in italics are part of the normal distribution of the data while the 7 points in italics are outliers from the normal distribution.

Comments

1. Looking at the final plot in figure 3.3(c) with all outliers removed, one can see it is a good straight line. Although it is possible to quantify how well the data are normally distributed, such calculations are not simple. Your ability to determine linearity by eye is actually very reliable and because of its simplicity this is the method most commonly used. The question of how we can statistically justify removing such outliers is the subject of the next section.
2. The one point from figure 3.3(b) that you may suspect is not an outlier is the point at 0.1177 M. If you recalculate the data with this point included and plot the Rankit plot it should be clear to you why this point was excluded.
3. If you go back to figure 1.3, which is a histogram of these same data, you will see that the identification of the outliers in the present example is thankfully consistent with this histogram where there are a number of analyses that appear to fall outside the normal distribution.

3.5 Test for an Outlier

An outlier is a value that does not belong to the distribution of the rest of the data. If it is included in calculating statistics such as the mean and standard deviation the estimates will not be representative of the true population mean and standard deviation. Therefore, it is important that outliers be identified and excluded from further calculations. In repeated chemical analyses, invariably outliers are found at the extremes (i.e., the biggest or smallest result). Remember, however, results cannot be simply discarded: there must be a basis for identifying data as outliers and a strategy for dealing with them. Outliers are still results and must be investigated and included in a report, even if they are not used in subsequent data analysis. When data have several outliers or contain values from two populations a graphical method such as a Rankit plot as in figure 3.3 is very useful in sorting out a normally distributed subset.

However, it is also useful to have a quick method to decide whether a particular value is an outlier or not. The method recommended by ISO is Grubbs's test, although many older texts still present Dixon's

Q-test for testing outliers. Using the mean, $\bar{x}$, and sample standard deviation, s, of the whole set, including the suspect outlier, $x_{suspect}$, the distance of the outlier from the mean is calculated as a number of standard deviations:

$$G = \frac{|x_{suspect} - \bar{x}|}{s} \tag{3.1}$$

G can be compared to tables of critical values for G at $\alpha = 0.05$, $G_{critical}$, calculated using equation 3.2 below. If $G > G_{critical}$ then the suspect point is rejected. Note that in the case of Grubbs's test, we compare with tabulated critical values simply because the calculation of the probability associated with the value of G is nontrivial. We do have a formula for the calculation of $G_{critical}$:

$$G_{critical} = \frac{(n-1)}{\sqrt{n}} \sqrt{\frac{t^2_{(0.05/n)''n-2}}{n - 2 + t^2_{(0.05/n)''n-2}}} \tag{3.2}$$

Table 3.2 Values of $G_{critical}$ used for Grubbs's test for outliers

$G_{critical}$	$G_{90\%}$	$G_{95\%}$	$G_{99\%}$	$G_{99.9\%}$
α	0.1	0.05	0.01	0.001
Number of data, n				
3	1.153	1.154	1.155	1.155
4	1.463	1.481	1.496	1.500
5	1.671	1.715	1.764	1.783
6	1.822	1.887	1.973	2.020
7	1.938	2.020	2.139	2.217
8	2.032	2.127	2.274	2.383
9	2.110	2.215	2.387	2.524
10	2.176	2.290	2.482	2.645
11	2.234	2.355	2.564	2.750
12	2.285	2.412	2.636	2.843
14	2.372	2.507	2.755	2.997
16	2.443	2.586	2.852	3.122
18	2.504	2.652	2.932	3.226
20	2.557	2.708	3.001	3.314
30	2.745	2.909	3.236	3.612
40	2.868	3.036	3.381	3.787
50	2.957	3.128	3.482	3.908

which may be implemented in Excel: =(n – 1)/SQRT(n)*SQRT((TINV(0.05/n, n – 2))^2/(n – 2+TINV(0.05/n, n – 2)^2)).

Values of $G_{critical}$ are given in table 3.2.

In other significance tests we can calculate a probability associated with the parameter; see the F test in example 3.4.

Example 3.2

The level of calcium in milk was determined using an EDTA titration method. Ten repeat measurements were performed with the following measured concentrations (units: mg g^{-1}): 4.59, 10.00, 6.07, 4.73, 9.91, 5.28, 16.65, 5.17, 4.59, and 4.38.

Problem

Determine whether there are any outliers in this data set.

Solution

1. The suspect values are 16.65 or 4.38—remember they have to be at the extremes of the data. Plot the data (figure 3.4) and it becomes obvious that if there is an outlier it has to

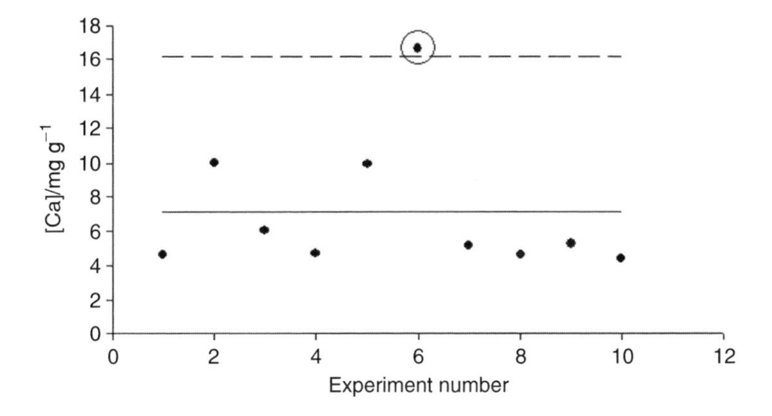

Figure 3.4 Data from 10 measurements of calcium in milk by EDTA titration. The outlier is circled. The solid line is the mean of the data ($\bar{x}$) and the dashed line is at $\bar{x} + G_{critical}s$, where s is the standard deviation of the data and $G_{critical} = 2.29$, which is the two-tailed G value at $\alpha = 0.05$.

be 16.65. Check 4.38 by all means, but there is no merit in statistically proving the obvious.

2. To solve this example you first need the mean and standard deviation. Use the Excel commands =AVERAGE(*range*) and =STDEV(*range*) which give 7.14 and 3.96 for the mean and standard deviation, respectively.
3. Calculate the absolute value of the standardized difference between the suspect value and the mean (i.e., the G statistic). This can be performed by hand or using Excel.

We will investigate the value 16.65 mg g^{-1} as the outlier by

$$G_{\text{suspect}} = \frac{|x_{\text{suspect}} - \bar{x}|}{s} = \frac{|16.65 - 7.137|}{3.961} = 2.402$$

The critical Grubbs's value for $\alpha = 0.05$ and $n = 10$ is 2.290.

As $G_{\text{suspect}} > G_{\text{critical}}$ for the value 16.65 mg g^{-1}, we reject H_0 (the null hypothesis is that the value is not an outlier) and we conclude that the point is an outlier. Another way of visualizing this is to calculate and plot the x-value that would just give G_{critical} by $x_{\text{critical}} = \bar{x} + sG_{\text{critical}}$. This is plotted as the dashed line in figure 3.4, and we see that the value of 16.65 mg g^{-1} is just greater than it.

These steps are illustrated for spreadsheet 3.3.

Spreadsheet 3.3

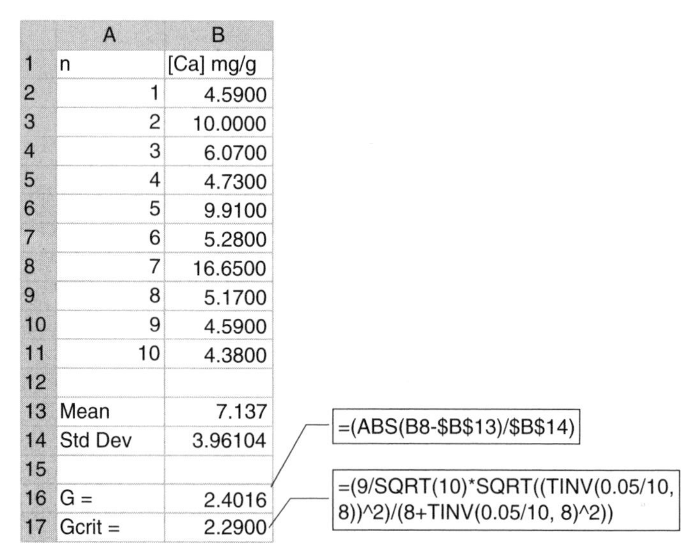

	A	B
1	n	[Ca] mg/g
2	1	4.5900
3	2	10.0000
4	3	6.0700
5	4	4.7300
6	5	9.9100
7	6	5.2800
8	7	16.6500
9	8	5.1700
10	9	4.5900
11	10	4.3800
12		
13	Mean	7.137
14	Std Dev	3.96104
15		
16	G =	2.4016
17	Gcrit =	2.2900

=(ABS(B8-$B$13)/$B$14)

=(9/SQRT(10)*SQRT((TINV(0.05/10, 8))^2)/(8+TINV(0.05/10, 8)^2))

Answer

As $G_{suspect} > G_{critical}$, the value 16.65 mg g^{-1} is an outlier and can be rejected at the 95% confidence level.

Comment

1. A query that may arise is if there are potential outliers at each end of the data what should we do? The answer is that Grubbs's test is only for a single outlier.
2. By discarding the outlier you are saying that it is not part of the normal distribution of data. Therefore, once it is rejected the mean and standard deviation can be recalculated, in this case to 6.08 and 2.25, respectively.
3. A common question is can you then perform an outlier test on the next furthest value from the mean? Ideally Grubbs's test is for one potential outlier only, although it is quite common to see the test then used on the next potential outlier. One must be careful, however, as if you start to reject too many points in a small data set (say less than 10 values) then it is likely that the data are not normally distributed. As all these statistics are based on the assumption of a normal distribution of data, then what you have been doing would be invalid. With large data sets (more than 10 values) the best approach to identifying outliers is to obtain a Rankit plot, where the points that deviate from the straight line are not part of the normal distribution.

It might not be surprising to learn that with only a few data a potential outlier to be excluded has to be quite far away from the other data. In many cases there is not enough information to reject H_0 (that the datum is not an outlier), although this does not mean that the suspect point is not an outlier. Remember that you only have as much information as the data can give you.

The next lesson is that an outlier is not necessarily the wrong answer, just one that is significantly different from the rest. Take as an example an international study of lithium in blood serum. Six laboratories took part and analyzed two samples having the same concentration of lithium (0.019 mM). Figure 3.5 shows that laboratory 4 appears to have quite different results from the other

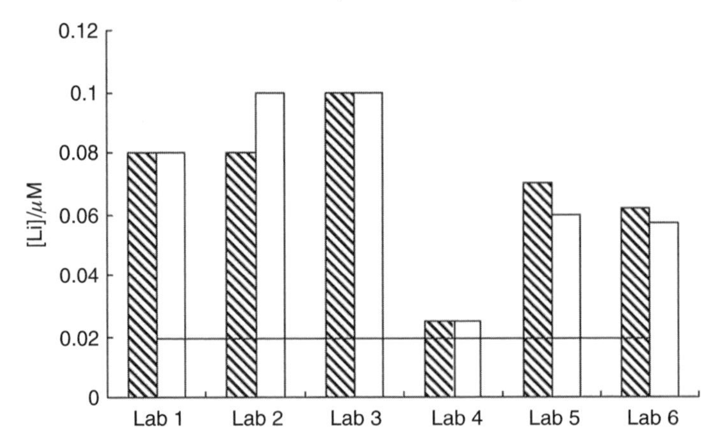

Figure 3.5 Results from a study of lithium in serum. Six laboratories analyzed a test portion of serum for Li in duplicate. The horizontal line is the accepted value. Data replotted from IMEP1. http://www.irmm.jrc.be/imep

laboratories. The G statistic for one result of 0.025 mM is 1.57, which is less than the 95% value of the two-tailed G distribution of 1.89. Even though the result of laboratory 4 looks very different from the others, it is not an outlier according to this statistical test, and this is just as well as it turned out to have the best answer.

Having decided that a datum should be rejected as an outlier, the mean and standard deviation of the data should be recalculated leaving out the errant result. However, always remember it can never be totally expunged.

3.6 Determining Significant Systematic Error

Systematic error in an analytical method must be determined and corrected for. We have seen that systematic error is assessed by making a measurement on a certified reference material (sometimes just referred to as a CRM). The mean of a number of determinations, $\bar{x}$, can be used to decide if the systematic error is significant by using the equation for a confidence interval of the mean

$$\bar{x} - \frac{t_{\alpha n-1}s}{\sqrt{n}} < x_{\mathrm{CRM}} = \mu < \bar{x} + \frac{t_{\alpha n-1}s}{\sqrt{n}} \qquad (3.3)$$

Taking x_{CRM} as the quantity value of the CRM (it is possible to extend the analysis if the value has its own uncertainty), then equation 3.3 can be used to calculate a t-value simply by setting $\mu = x_{CRM}$:

$$x_{CRM} = \bar{x} \pm \frac{t_{\alpha n-1} s}{\sqrt{n}} \tag{3.4}$$

and rearranging to give

$$t = \frac{|\mu - \bar{x}|\sqrt{n}}{s} = \frac{|x_{CRM} - \bar{x}|\sqrt{n}}{s} \tag{3.5}$$

With the test for systematic error the null hypothesis is that there is no systematic error, that is, the mean of the population of measurement results, μ, is x_{CRM}. The associated probability with the t-value found using equation 3.5 can be calculated in Excel by TDIST (t, $n-1$, 2). This probability is the fraction of repeated analyses that would have the observed $|x_{CRM} - \bar{x}|$ or greater from a measurement that really has no systematic error. The Excel function TDIST (t, $n-1$, *tails*) calculates the probability of a t-value at $n-1$ degrees of freedom. The parameter *tails* takes the value 1 or 2 (see section 2.2). If it is known that the error can be *only* positive or *only* negative, then all the probability can be considered at that end of the distribution and *tails* = 1. Usually there is no particular reason why the systematic error is either positive or negative and so *tails* = 2. Figure 3.6 illustrates the distribution of the probability in each of these cases.

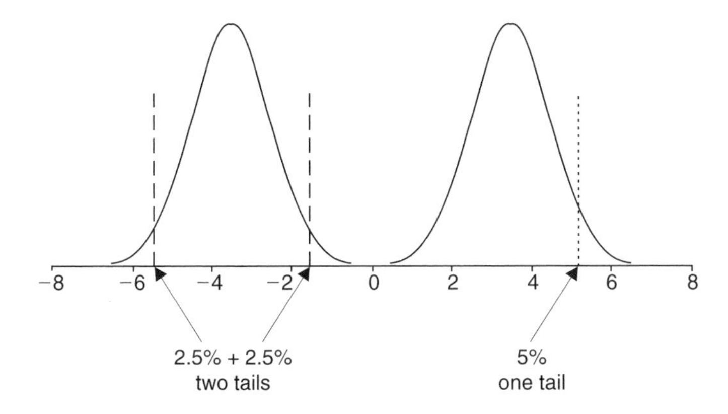

Figure 3.6 One- and two-tailed probabilities. The extreme 5% of the distribution can be shared equally between each end (two tailed), or the entire 5% can be located at one end or the other.

As in all such testing if the probability falls below the limit decided ($p < \alpha$) then the null hypothesis, that there is no systematic error, is rejected, and we conclude that there *is* systematic error.

Example 3.3

The fluoride content in toothpaste is measured using a fluoride ion-selective electrode. To perform the measurement requires the sample to be prepared by extraction of the fluoride from the toothpaste. To determine whether the extraction and measurement procedure are free of systematic error a number of analysts measured the fluoride content of a test toothpaste sample with assigned mass fraction of 0.033 m/m%. The measured mass fractions of fluoride expressed as a percentage determined by the nine analysts were 0.042, 0.040, 0.028, 0.035, 0.044, 0.035, 0.041, 0.043, 0.040 m/m%.

Problem

Determine whether there is a systematic error in the method with 95% probability.

Solution

Using Excel:

1. Plot the data (figure 3.7). It looks as if there might be an outlier with value 0.028 m/m%.
2. Calculate the mean and standard deviation of the nine test results.
3. Test for any outliers using Grubbs's test: entering = ABS((*x* – AVERAGE(*range*))/STDEV(*range*)) to calculate G and compare with $G_{critical}$. In this case $G = 2.092$ while $G_{critical} = 2.215$ for 9 data and the point is retained.
4. Calculate the value of t, $t = (|x_{assigned} - \bar{x}|\sqrt{n})/s$, using = ABS (*assigned value* – AVERAGE(*range*))*SQRT(*n*)/STDEV (*range*). For the data in spreadsheet 3.4 this t-value is 3.334.
5. Calculate the probability of this t-value using TDIST(*t*, *n* – 1, *tails*), which gives $p = 0.0103$.

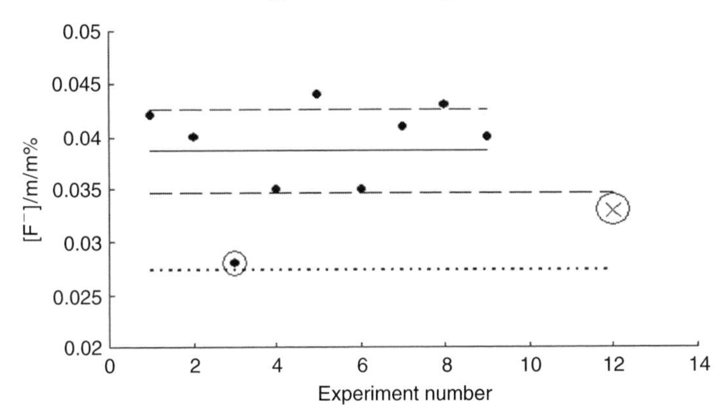

Figure 3.7 Results from the measurement of fluoride in toothpaste. Data 1–9 are replicate measurements. Dashed lines are ±95% confidence intervals. The circled datum (number 3) is the suspect outlier. The dotted line is the critical x value calculated from the $\alpha = 0.05''$, $n = 9$ Grubbs's test. The datum numbered 12 (circled cross) is the reference value of the sample (0.033 m/m%).

Spreadsheet 3.4

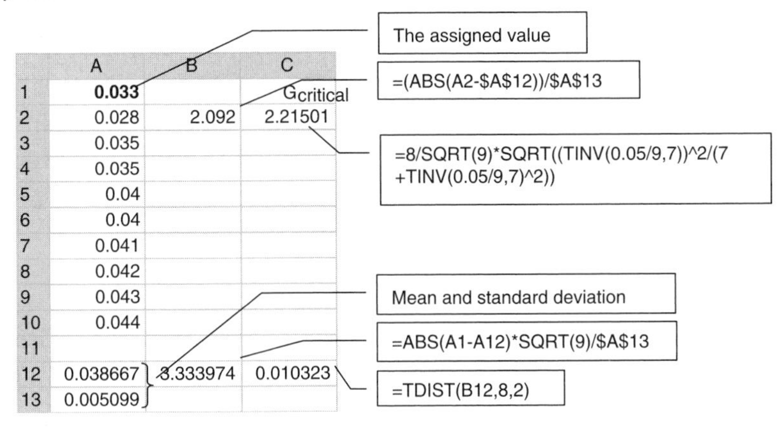

	A	B	C
1	**0.033**		Gcritical
2	0.028	2.092	2.21501
3	0.035		
4	0.035		
5	0.04		
6	0.04		
7	0.041		
8	0.042		
9	0.043		
10	0.044		
11			
12	0.038667	3.333974	0.010323
13	0.005099		

These steps are illustrated for spreadsheet 3.4, where cell A1 contains the assigned value.

Answer

The operations in cells B12 and C12 to test for systematic error show the probability of H_0 (that there is no systematic error) is 0.010323. Therefore as $p < 0.05$ the null hypothesis is rejected at the 95% level and the conclusion is made that there is a systematic error in the ion-selective electrode method.

Comment

1. If you inspected the data you might suspect there was a systematic error, as all but one of the data are higher than the assigned value; however, without performing the test you could not be confident that this was the case. Plotting the points makes you even more concerned that 0.028 m/m% is an outlier. However, it falls just inside the critical value and cannot be rejected.
2. The assigned value lies just outside the 95% confidence interval which allows us to conclude there is a significant difference. Note that plotting the data always is useful, but you must calculate the probabilities.
3. Having determined $p = 0.0103$ we can conclude that H_0 is just accepted with 99% probability.

It is instructive to consider equation 3.5. If the sample mean ($\bar{x}$) just happens to coincide with μ then t is zero and there is no question that the null hypothesis is supported by the data and the conclusion is that there is no systematic error. The larger the value of t the more likely you are to reject the null hypothesis and conclude there is a systematic error. Equation 3.5 shows there are three key factors that will give a larger value of t. The first is that the greater the difference between the sample mean and the assigned population mean, the greater the value of t. The second is that t increases as the number of repeats increases; this is because as more samples are measured the confidence in the sample mean increases. The third factor that increases t is a decrease in the standard deviation s. What these three factors mean is the more care the analyst takes by performing repeated measurements with a lower overall standard deviation the greater the value of t and hence the greater the likelihood of a systematic error being declared. This may at first seem unfair, as it suggests the sloppier the chemist the less likely there is of a systematic error. In reality it means the more careful chemist is more able to identify any systematic error in the analysis. It may be better to think of a t-test that arrives at the conclusion that H_0 should *not* be rejected as showing not that H_0 is true, but that there is insufficient evidence to reject H_0: in legal terms "not proven" rather than "not guilty."

3.7 Testing Variances: Are Two Variances Equivalent?

When choosing an analytical method to analyze a sample you need to consider a number of things including the level of precision required. This is done by comparing the variances of the analytical methods. To decide if there is a significant difference between variances the probability associated with the Fisher F-statistic is calculated. Given two standard deviations s_1 and s_2, where $s_1 > s_2$

$$F = \frac{s_1^2}{s_2^2} \tag{3.6}$$

For identical standard deviations $F = 1$. There is a known distribution of F given the degrees of freedom of s_1 and s_2, which allows calculation of the fraction of measurements of s_1 and s_2 from populations of equal σ that will lead to an F-value equal to or greater than the value determined by equation 3.6. In terms of significance testing, the null hypothesis is that the population standard deviations are equal, and we decide there is a significant difference if the probability of F falls below, say, 0.05 (95%). In Excel the probability is given by =FDIST(F, $n_1 - 1$, $n_2 - 1$). Calculation of the F-statistic is at the heart of analysis of variance (ANOVA), and is also used to check for equality of variance before undertaking t-tests of means (see section 3.8).

Example 3.4

Problem

An analytical laboratory performs the analysis of copper in tap water by extracting the copper into chloroform that contains the chelating ligand diethyldithiocarbamate (DEDTC). The resultant $Cu(DEDTC)_2$ complex is yellow and can be monitored at 436 nm using a UV–visible spectrophotometer. The manager of the laboratory wants to know whether there is a significant difference between the three analysts who perform this analysis with regards to their precision. Each analyst performs nine analyses of a reference material of assigned concentration of 1.6 ppm Cu^{2+}. The results of each analyst are shown in table 3.3.

Table 3.3 Replicate results from three analysts for the measurement of the mass fraction of copper in a solution of nominal concentration of 1.6 ppm

Analyst 1 (ppm)	*Analyst 2 (ppm)*	*Analyst 3 (ppm)*
1.61	1.55	1.68
1.48	0.73	1.05
1.71	1.52	1.52
1.48	1.56	1.14
1.53	1.64	1.67
1.57	1.60	2.11
1.78	1.61	2.16
1.52	1.84	2.20
1.84	1.49	0.95

An F-test is done to ascertain whether there is a significant difference, at the 95% level, in the standard deviations obtained by pairs of analysts.

Solution

This is most easily performed using Excel. The following steps need to be performed:

1. Calculate the standard deviation of the results of each analyst.
2. From the standard deviations calculate the F-value for each combination of two analysts using equation 3.6. The important thing to remember is that the numerator is always the larger standard deviation.
3. Calculate the probability associated with the F-values using =FDIST(F, $n_1 - 1$, $n_2 - 1$). If the probability is greater than 0.05 the null hypothesis holds and it is concluded that there is no significant difference. If $p < 0.05$ the null hypothesis is rejected and it is concluded that there is a significant difference.

These steps are illustrated for spreadsheet 3.5.

Do not forget to square the standard deviations! We compare variances.

Spreadsheet 3.5

	A	B	C
1	Analyst 1	Analyst 2	Analyst 3
2	1.61	1.55	1.68
3	1.48	0.73	1.05
4	1.71	1.52	1.52
5	1.48	1.56	1.14
6	1.53	1.64	1.67
7	1.57	1.6	2.11
8	1.78	1.61	2.16
9	1.52	1.84	2.2
10	1.84	1.49	0.95
11			
12			
13			
14			
15	0.133041	0.307535	0.485707
16			
17	F value		
18	2 versus 1	3 versus 2	3 versus 1
19	5.3433773	2.4943609	13.328311
20	0.014464	0.1088231	0.0006992

Standard deviation

=(C15^2/A15^2)

=FDIST(C19, 8, 8)

Answer

From the probabilities shown in cells A20, B20, and C20 it can be seen that there is a significant difference at the 99.93% level between analysts 1 and 3 and at the 98.6% level between analysts 1 and 2, but there is not a significant difference (i.e., $p > 0.05$) between analysts 2 and 3.

Comments

1. Thus it appears that analyst 1 has a problem with his or her precision. However, it should be pointed out that in fact analyst 1 has a much smaller standard deviation than analysts 2 and 3 and it is these two that could be in error. How the failing of the null hypothesis between analysts 1 and 2 is interpreted is not the province of statistics and needs to be assessed within the laboratory.
2. When calculating an F-value one common question is of the two standard deviations which is s_1 and which is s_2? The answer is simply the data set with the largest standard deviation is s_1. What this means is F is always greater than 1.

3.8 Testing Two Means (Means t-Test)

A common problem is to compare two or more sets of data. For example, a new analytical method may be assessed by analyzing a test material using the new and an established method. The means of a number of replicate measurement results obtained by each method will not be identical, but within the experimental uncertainty is there a significant difference? The preferred method is to use analysis of variance (ANOVA) which can accommodate a number of variables and different numbers of data (see chapter 4). However, a quick test for two means can be performed by calculating a t-statistic and its associated probability.

First it is necessary to determine if the standard deviations of the samples are significantly different. This is done by calculating the F-statistic, as described above.

If the standard deviations are not significantly different then a pooled standard deviation is calculated:

$$s_p = \sqrt{\frac{(n_1 - 1)s_1^2 + (n_2 - 1)s_2^2}{n_1 + n_2 - 2}} \qquad (3.7)$$

with $n_1 - 1 + n_2 - 1 = n_1 + n_2 - 2$ degrees of freedom, and equation 3.3 can be employed once again to give

$$t = \frac{|\bar{x}_1 - \bar{x}_2|}{s_p\sqrt{1/n_1 + 1/n_2}} \qquad (3.8)$$

for which the probability may be determined. $\bar{x}_1$ and $\bar{x}_2$ are the means of n_1 and n_2 data.

If the standard deviations are significantly different as tested by the F-statistic, the t-value is now calculated by

$$t = \frac{|\bar{x}_1 - \bar{x}_2|}{\sqrt{s_1^2/n_1 + s_2^2/n_2}} \qquad (3.9)$$

with degrees of freedom calculated from

$$df = \frac{\left(s_1^2/n_1 + s_2^2/n_2\right)^2}{\left[(s_1^4/n_1^2(n_1 - 1)) + (s_2^4/n_2^2(n_2 - 1))\right]} \qquad (3.10)$$

and then rounding down to the nearest integer. (Note that the Excel function =ROUND(*x*, *decimal places*) rounds up 5–9 and down 0–4, but it is possible to force rounding down by =ROUNDDOWN (*x*, *decimal places*), which is considered a more conservative approach.)

In each case the null hypothesis is that the two samples come from populations with equal means (i.e., that $\mu_1 = \mu_2$, not that $\bar{x}_1 = \bar{x}_2$, which is clearly nonsense). As the *t*-value increases the probability of the null hypothesis decreases, and when the probability reaches a suitably low value ($p < \alpha$), H_0 can be rejected. Note that the assumption of unequal variances leads to a higher probability of the data given H_0, and so is more conservative in rejecting H_0. For this reason, some authorities recommend it best to always assume unequal variances.

Example 3.5

An analytical laboratory analyses the glucose levels in soft drinks using a spectroscopic enzyme assay and is considering using an enzyme electrode instead. To ascertain whether the spectroscopic assay and the enzyme electrode give means that are not significantly different, a soft drink was analyzed six times by the same analyst using each method. The concentration of glucose (units: mM) determined for the 12 test portions were:

- Using the spectroscopic assay: 1.90, 1.82, 1.70, 1.94, 1.85, 1.90
- Using the enzyme electrode: 1.35, 1.65, 1.76, 1.41, 1.80, 1.33

Problem

At what probability are the mean results obtained by each method significantly different? Would you infer that the methods did indeed give different results?

Solution

The solution to this problem is to test the hypothesis that the two methods have equal population means. This is therefore a

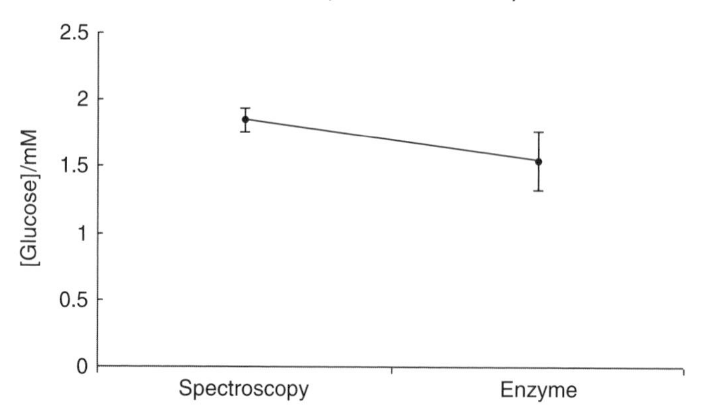

Figure 3.8 Means and 95% confidence intervals on the analysis of a sample of glucose by a spectrophotometric method and an enzyme method.

problem of testing two means. The steps involved in testing the two means are:

1. Calculate the mean and standard deviation for each set of data using =AVERAGE(*range*) and =STDEV(*range*). For the spectroscopic assay the values of the mean and standard deviation are 1.852 and 0.085 mM, respectively, and for the enzyme electrode they are 1.550 and 0.212 mM, respectively. These are plotted for comparison in figure 3.8.
2. Calculate the *F*-statistic from s_1^2/s_2^2, where s_1 is the larger standard deviation, in this case 0.212 mM for the enzyme electrode ($F = 6.16$).
3. Determine the probability associated with this *F*-statistic by entering =FDIST(F, $n_1 - 1$, $n_2 - 1$), where n_1 and n_2 are the number of data (here both are equal to 6). From spreadsheet 3.6 for these data we see $p = 0.034$ and therefore we conclude that the standard deviations are significantly different at the 96.4% level. As a consequence, equations 3.9 and 3.10 should be used to calculate the *t*-statistic and the degrees of freedom.
4. Calculate the *t*-statistic using equation 3.9 which in Excel is =ABS(mean$_1$ – mean$_2$)/SQRT(S_1^2/n_1 + S_2^2/n_2). Here $t = 3.234$ where $n_1 = 6$ and $n_2 = 6$.
5. Calculate the degrees of freedom using equation 3.10 which in Excel is =(s_1^2/n_1 + s_2^2/n_2)^2/(s_1^4/(n$_1$^2*(n$_1$ – 1)) + s_2^4/(n_2^2*(n$_2$ – 1))). This function gives 6.583, so round

Spreadsheet 3.6

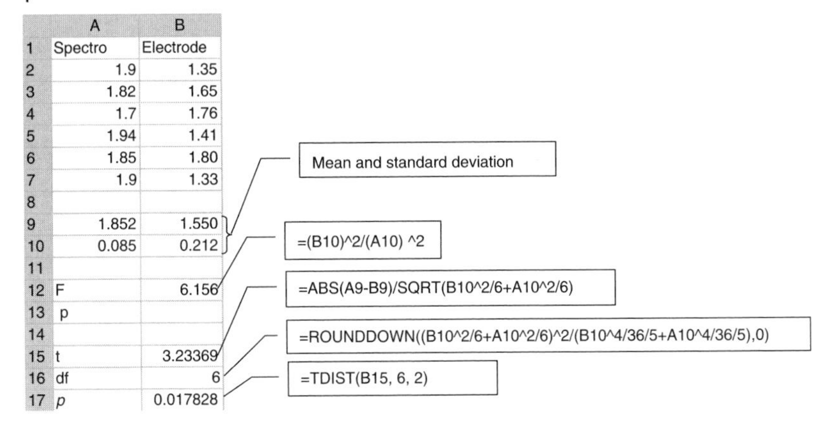

	A	B
1	Spectro	Electrode
2	1.9	1.35
3	1.82	1.65
4	1.7	1.76
5	1.94	1.41
6	1.85	1.80
7	1.9	1.33
8		
9	1.852	1.550
10	0.085	0.212
11		
12	F	6.156
13	p	
14		
15	t	3.23369
16	df	6
17	*p*	0.017828

down to a result of 6 degrees of freedom (or use ROUNDDOWN directly).

6. Calculate the associated probability for the t-statistic using TDIST(*t*, *df*, *tails*) which gives the answer $p = 0.017828$ (see spreadsheet 3.6).

Answer

Hence, the two data sets are concluded to come from populations with different means ($p = 0.018$, or with 98.2% probability).

Comments

1. If we had been asked to test at the 95% probability level the two analytical methods would be judged to give different results, but if the question were asked at 99% probability then we would conclude the null hypothesis was accepted and the difference between the means is not considered significant.
2. There was a significant difference in the standard deviations at the 95% level ($p < 0.05$) and therefore equations 3.9 and 3.10 were used to calculate the t-statistic and its associated degrees of freedom. If there were no significant difference between the standard deviations then the pooled standard deviation and the t-statistic could be calculated using equations 3.7 and 3.8.

3. As with the other statistical tests, rather than compare to a table of values of the t-statistic, when possible we use Excel to calculate the probability associated with a t-statistic. This example emphasizes that the question of significance is a subjective one. You choose the probability at which you will reject H_0.
4. The statement you can make about this process is the following: "If the experiments were repeated a large number of times under exactly the same conditions with the same global mean result, the calculated t-statistic for the difference of a pair of means of six replicates by each method would equal or exceed 3.23 in 1.8% of the cases." "There is a significant difference with 98.2% probability" is shorter, but remember what the true statement is.

3.9 Paired *t*-Test

In some cases we do not have the luxury of repeated measurements of a single test material, but do have one-off measurements of a number of different test materials performed by two methods. The two methods can be compared by considering the results of each pair of one-off measurements. This is possible as for a particular test material measured by each method the *difference* in the result should be zero if the two methods give equivalent results. For a number of analyses of different materials any pair of materials is the same and so the mean of the differences can be tested against zero. If the two methods give equivalent results within measurement uncertainty the difference between results on the same material by each method should be zero. In a paired t-test, therefore, the mean $\bar{x}_\mathrm{d}$ and standard deviation s_d of the differences are calculated and a t-statistic determined from equation 3.5 with $\mu = 0$:

$$t = \frac{|\bar{x}_d|\sqrt{n}}{s_d} \qquad (3.11)$$

where n is the number of differences, that is, the number of pairs of results. H_0 is that the population mean of the differences is zero, which is the case if the two methods give results with equal population means.

Table 3.4 Replicate results for the measurement of calcium in milk by two methods

Test material	*1*	*2*	*3*	*4*	*5*	*6*	*7*	*8*	*9*
AAS[a] (mg g^{-1})	3.01	2.58	2.52	1.00	1.81	2.83	2.13	5.14	3.20
CT[b] (mg g^{-1})	2.81	3.20	3.20	3.20	3.35	3.86	3.88	4.13	4.86

[a] Atomic absorption spectroscopy.
[b] Complexometric titration.

Example 3.6

The amount of calcium in different samples of milk powder (in mg of calcium per g of milk powder) were analyzed by two methods, one employing extraction followed by analysis using atomic absorption spectroscopy, the other using a complexometric titration method. The results of nine analyses are shown in table 3.4.

Problem

Determine whether the two analytical methods give equivalent analytical results.

Solution

As we have a number of one-off measurements, we will use a paired *t*-test.

1. For each pair of data the difference is calculated (see spreadsheet 3.7 and figure 3.9).
2. Calculate the average and standard deviation of the differences (−0.92 mg g^{-1} and 1.024 mg g^{-1} in this case).
3. Calculate the *t*-value from the mean and standard deviation using equation 3.11: $t = (|\bar{x}_d|\sqrt{n})/s_d = (0.92 \times \sqrt{9})/1.024 = 2.6905$.
4. From the *t*-value calculate a probability using =TDIST (*t*, *n* − 1, *tails*), where n is the number of pairs, which for this set of data is =TDIST(2.6905, 8, 2), see spreadsheet 3.7. $p = 0.0275$ means if we were to reject the null hypothesis we would do so knowing we would make an error in 2.75% of repeated experiments. As this is less than the usual 5% given

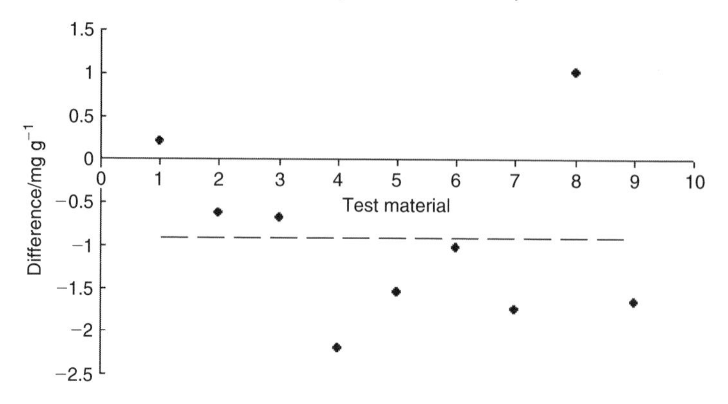

Figure 3.9 Difference between measurements of a number of test samples of milk powder for calcium by two methods. The dashed line is the mean of the differences.

Spreadsheet 3.7

	A	B	C
1	AAS	CT	Difference
2	3.01	2.81	0.20
3	2.58	3.20	-0.62
4	2.52	3.20	-0.68
5	1.00	3.20	-2.20
6	1.81	3.35	-1.54
7	2.83	3.86	-1.03
8	2.13	3.88	-1.75
9	5.14	4.13	1.01
10	3.20	4.86	-1.66
11			
12			-0.92
13			1.024591
14			
15		t value	2.690505
16		p	0.027475

=(A3-B3)

Means and standard deviation of differences

=(ABS(C12)*SQRT(10))/C13

=TDIST(C15, 8, 2)

by a test at 95% confidence we may decide that there is sufficient evidence to reject the null hypothesis and conclude there is a significant difference between the results of the two methods.

Answer

The probability of the data given that there is no difference between the methods is 0.0275. Therefore we conclude that there is a significant difference in the analytical result obtained with the AAS method of analysis of calcium in milk and the complexometric titration method.

Comment

1. In seven out of the nine pairs the AAS method gave a smaller value than the complexometric method, which should rouse the suspicions of the analyst.
2. Without measurements on a CRM we cannot say whether either of the two methods give a result that is without bias.

3.10 Hypothesis Testing in Excel

The equations above can be used to determine t- and F-statistics using TDIST and FDIST to give the probability. Excel also provides a function TTEST(*range1,range2,tails,type*) where *type* is 1 for a paired t-test, 2 for a means test with equal variances, and 3 for a means test with unequal variances, and *tails* is 1 or 2. The output of this function is the probability of the data given the null hypothesis. There are also menu-driven calculations in the Analysis Toolbox. Choose Tools, Data Analysis from the menus, and there are three items: t-Test: Paired Two Sample for Means; t-Test: Two Sample Assuming Equal Variances; and t-Test: Two Sample Assuming Unequal Variances. The dialogue boxes that appear when one of these items is chosen require the two data ranges, the hypothesized mean difference (usually zero, but it must be entered), whether there are column labels, the probability level for a test (default 0.05), and an output range. The output includes the one- and two-tailed probability values and the t-statistic at the probability level for the test. Spreadsheet 3.8 shows the box for t-Test: Two Sample Assuming Equal Variances.

Spreadsheet 3.8

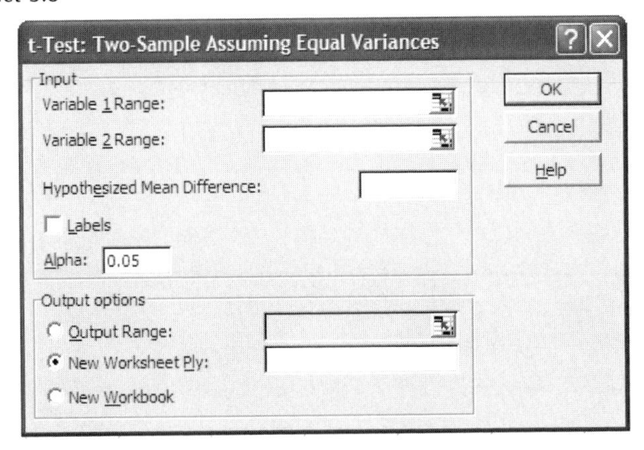

4

Analysis of Variance

4.1 What This Chapter Should Teach You

- What ANOVA is, and what it is used for.
- To perform and interpret a one-way ANOVA.
- To determine which effects are significant using least significant difference.
- To perform and interpret a two-way ANOVA.

4.2 What Is Analysis of Variance (ANOVA)?

ANOVA is the workhorse method of using statistics to compare means and determine the effects of influence factors on measurement results (i.e., anything that can be varied or measured that may affect the result). In chapter 3 we learned how to use Student t-tests to compare two means. There is nothing to stop us performing a series of t-tests on pairs of means that must be compared, but a different approach that looks at the variance of data, ANOVA, can decide if there is a significant effect caused by a factor for which we have any number of sets of data. ANOVA relies on an understanding of two things. First, how the variances of different components can be combined to give the overall observed variance of data. Second, that a difference in means can lead to a spread of results of the combined data that can be detected in terms of an increased variance. As an example, consider an attempt to determine if there is a significant difference between the means of replicate analyses conducted by two methods. The standard deviation of each set of results will estimate the repeatability of the measurement. If the two methods have different means then the standard deviation of the combined data will

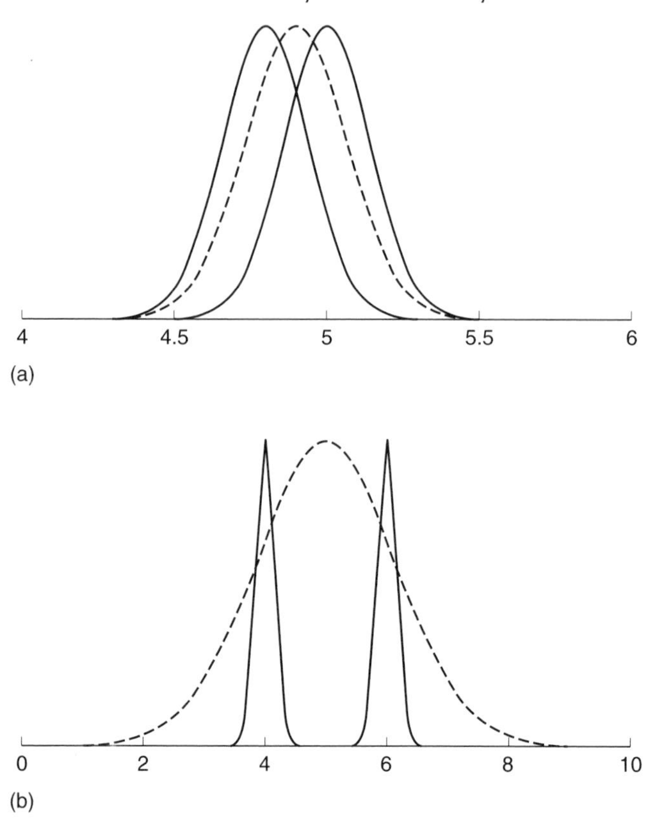

Figure 4.1 Effect of a difference in means on the standard deviation of data: (a) two normal distributions with $\mu_1 = 4.8$, $\sigma_1 = 0.14$ and $\mu_2 = 5.2$, $\sigma_2 = 0.14$ and the distribution of their combination, $\sigma = 0.16$ (dashed line); (b) two normal distributions with $\mu_1 = 3.9$, $\sigma_1 = 0.14$ and $\mu_2 = 5.9$, $\sigma_2 = 0.14$ and the distribution of their combination, $\sigma = 1.2$ (dashed line).

be increased by any differences arising from the methods. This is illustrated in figure 4.1. When the means are far apart, even though the individual standard deviations are not great, the combination has a huge standard deviation. ANOVA is powerful because it can determine if there is significant difference among a number of instances of the same factor (e.g., if we wanted to know if there were any difference in the result between three or more analytical methods), and also among different factors (e.g., what is the effect of temperature and concentration on the yield of a reaction?). ANOVA allows us to obtain a probability of finding the observed data given that there is no effect of a particular influence factor (the null hypothesis in ANOVA tests). Following a significance test it is then possible to apportion the variance among the significant effects.

ANOVA does not require the effects to be independent, but the data do need to be normally distributed.

In this chapter we briefly show how an ANOVA is performed for the simplest case of a single factor (so-called one-way ANOVA) and then for a two-way ANOVA. ANOVA is available (albeit in a restricted form) in Excel, and in most other statistical packages. Although we shall show you how to do a one-way ANOVA by hand, the chapter will concentrate on the interpretation of ANOVA output from software applications.

4.3 Jargon

We use the word "factor" to refer to the quantity that is being investigated. An example would be an investigation of changing the solvent polarity on the result of a high-performance liquid chromatography (HPLC) analysis. The factor is solvent polarity and we may make measurements, in the example, using three solvents of different polarity. The ANOVA we do in this case is a single-factor or one-way ANOVA because there is only one factor being investigated—the solvent polarity. In an example later in the chapter, two factors are considered in the calibration of pipettes, the method of using the pipette and the analyst. A factor may be a continuous variable, such as temperature, or may be discrete entities such as analysts, or drums of chemicals, or pipettes.

How much the measurand changes as the factor is varied is known as the "effect" of the factor. Often in ANOVA we are only interested in testing whether there is any effect at all. In this case we use the methods of significance testing explained in chapter 3 and test the null hypothesis that the observed variance arises from random effects. If the hypothesis is rejected at a particular probability (say 95%) then we conclude that the effect is significant.

4.4 One-Way ANOVA

In a one-way ANOVA we have instances of the factor being investigated, with replicate results for each instance. For example we may be evaluating the performance of three laboratories that have each been asked to perform duplicate determinations of identical samples. Thus here the factor is "laboratory" of which we have three instances—laboratory A, laboratory B, and laboratory C. The data (each measurement result) are laid out in a matrix with the instances of the factor in each column and the replicates in each row (table 4.1).

In our example there will be three columns, one for each laboratory, and two rows for the duplicate measurement results. This is shown in table 4.2.

The steps in calculating the ANOVA are as follows:

1. Calculate the mean of the entire data ($\bar{\bar{x}}$):

$$\bar{\bar{x}} = \frac{\sum_j \sum_i x_{i,j}}{N}$$

   where $N = \sum_j n_j$ is the total number of measurement results. This mean is called the grand mean. Subtract $\bar{\bar{x}}$ from each measurement result, that is, $x_{i,j}$ becomes $(x_{i,j} - \bar{\bar{x}})$. This is

Table 4.1 General layout of data for a one-way ANOVA

	Instances of the Factor j = (1 . . . k)			
	1	*2*	. . .	*k*
Replicates, $i = (1 \ldots n_j)$				
1	$x_{1,1}$	$x_{1,2}$	. . .	$x_{1,k}$
2	$x_{2,1}$	$x_{2,2}$	. . .	$x_{2,k}$
⋮	. . .	. . .	$x_{i,j}$	. . .
⋮	. . .	. . .	. . .	. . .
n_j	$x_{n_1,1}$	$x_{n_2,2}$	. . .	$x_{n_k,k}$

Table 4.2 Data entry grid for an ANOVA of the results of an interlaboratory study

	Laboratory		
Replicate	*A*	*B*	*C*
1			
2			

known as correction for the mean and the values are called the mean-corrected values.

2. Square each mean-corrected value and then sum them all to give the total sum of squares also known as the corrected sum of squares:

$$SS_{\mathrm{T}} = \sum_{i=1}^{i=n_j} \sum_{j=1}^{j=k} (x_{i,j} - \bar{\bar{x}})^2$$

3. For each column (i.e., instance of the factor) average the mean-corrected values

$$= \sum\nolimits_{i=1}^{i=n_j} \frac{(x_{i,j} - \bar{\bar{x}})}{n_j}$$

square this average

$$= \left(\sum_{i=1}^{i=n_j} \left(\frac{x_{i,j} - \bar{\bar{x}}}{n_j} \right) \right)^2$$

and multiply by the number of rows (n_j) for that column

$$= n_j \left(\sum_{i=1}^{i=n_j} \left(\frac{x_{i,j} - \bar{\bar{x}}}{n_j} \right) \right)^2$$

Sum across the columns (sum $j = 1$ to k) to give the sum of squares due to the factor studied:

$$SS_{\mathrm{c}} = \sum_{j=1}^{j=k} n_j \left(\sum_{i=1}^{i=n_j} \left(\frac{x_{i,j} - \bar{\bar{x}}}{n_j} \right) \right)^2$$

SS_{c} is also known as the treatment sum of squares, heterogeneity sum of squares, or the between column sum of squares. SS_{c} is related to the variance between factors.

4. Calculate the residual sum of squares as $SS_{\mathrm{r}} = SS_{\mathrm{T}} - SS_{\mathrm{c}}$. SS_{r} is also known as the within variables sum of squares.

Having obtained these values, they are laid out in an ANOVA table (table 4.3).

Table 4.3 The output of a one-way ANOVA

Source	*Sum of Squares*	*Degrees of Freedom*	*Mean Square*	*F*
Between factors	$SS_c = \sum_{j=1}^{j=k} n_j \left(\sum_{i=1}^{i=n_j} (x_{i,j} - \bar{\bar{x}})/n_j \right)^2$	$k-1$	$\overline{SS}_c = \frac{SS_c}{k-1}$	$F = \frac{\overline{SS}_c}{\overline{SS}_r}$
Within factor	$SS_r = SS_T - SS_c$	$N-k$	$\overline{SS}_r = \frac{SS_r}{N-k}$	
TOTAL	$SS_T = \sum_{i=1}^{i=n_j} \sum_{j=1}^{j=k} (x_{i,j} - \bar{\bar{x}})^2$	$N-1$		

N is the total number of data, k the number of instances of the factor, and $\bar{\bar{x}}$ the grand mean.

Note that if there are the same number of repeats for each instance of the factor ($n_j = n$ for all j), then $N = kn$ and the within factor degrees of freedom $= k(n-1)$. The residual mean square ($\overline{SS}_r$) is an estimate of the average variance of the results within each instance of the factor. In the case of replicated analytical results $\overline{SS}_r$ is s_r^2, an estimate of the repeatability variance σ_r^2 (see section 2.7). ANOVA is therefore a useful way of estimating measurement precision. We usually want to know something about the differences among our factor: for example, "Does laboratory A get the same results as laboratory B?" Unfortunately, $\overline{SS}_c$ does not tell us the answer on its own because it includes the measurement variance as well as anything to do with the differences between the instances of the factor. Any measurement will have an uncertainty so there is no simple way of teasing out the pure effect of the factor being studied. Testing for a significant effect requires us to compare the between factors and within factor mean squares. If there were no effect, $\overline{SS}_c$ and $\overline{SS}_r$ would be equal. Therefore we want to know whether the differences between the instances of the factor are significantly greater than what we would expect from the measurement repeatability, or "is $\overline{SS}_c$ significantly greater than $\overline{SS}_r$?" This is accomplished by determining the F-statistic (see chapter 3) as the ratio of these quantities and finding its probability at the requisite degrees of freedom ($k-1$ for $\overline{SS}_c$, and $N-k$ for $\overline{SS}_r$). The probability is one tailed because this is an example where $\overline{SS}_c$ cannot be smaller than $\overline{SS}_r$, so we only need to compute the probability that it is significantly greater.

4.4.1 Calculating the standard deviation due to the factor studied

It may be that the objective is only to determine if there is a significant effect, but for some problems estimates of the within factor standard deviation and between factors standard deviation are required. An example is in sampling studies where the standard deviation of the results is known to be composed of a measurement standard deviation and the heterogeneity of the test material itself, both of which must be estimated. For the case of equal numbers of replicate measurements (i.e., $n_j = n$ for all j) the between factors mean square $\overline{SS}_c$ is an estimate of $\sigma_r^2 + n\sigma_c^2$, where σ_c^2 is the variance of the effect of the factor and σ_r^2 the repeatability variance:

$$\overline{SS}_r = s_r^2 \approx \sigma_r^2, \ \overline{SS}_c = s_c^2 \approx \sigma_r^2 + n\sigma_c^2 \tag{4.1}$$

therefore

$$\sigma_r \approx s_r = \sqrt{\overline{SS}_r} \tag{4.2}$$

$$\sigma_c \approx s_c = \sqrt{\frac{\overline{SS}_c - \overline{SS}_r}{n}} \tag{4.3}$$

Hence from ANOVA we have estimates for the repeatability standard deviation (s_r) and the standard deviation associated with the effect (s_c).

In an ANOVA in which a number of replicate analytical measurements have been made we can use equation 4.2 to calculate the repeatability and equation 4.3 to calculate the standard deviation due to the factor.

4.5 Least Significant Difference

Although ANOVA can determine if there is a significant effect of a factor, if there are more than two instances of the factor then when a significant difference is found ANOVA does not tell which of those instances contributes significantly to the difference. For example, suppose we use ANOVA to decide if there is a significant difference among the delivered volume of a batch of five 10 mL pipettes, by weighing ten deliveries of each pipette. The data matrix has five columns (each pipette) and ten rows (repeats), hence a total of 50 data. If the ANOVA does conclude there is a significant difference, was there just one pipette that was different from the other four, or were there more? It would be possible to find out by testing each against the nominal 10 mL by a *t*-test, or each against another in a series of means *t*-tests. A reasonable idea about which pipette or pipettes are different may be determined by the method of least significant difference (LSD). The ANOVA has given the within factor standard deviation ($\sqrt{SS_r}$) which in this case is the repeatability of the delivery and weighing (s_r). The difference between the means of any two columns of data would be expected to have a standard deviation of $\sqrt{2}s_r/\sqrt{n}$, where n is the number of replicates, here 10 (see section 2.7). The 95% confidence interval of the difference of any two means is $\sqrt{2}t_{0.05'',45}s_r/\sqrt{10}$ in this example, where the *t*-value is obtained at 45 degrees of freedom ($N-k=50-5=45$). The degrees of freedom is 45 since there are a total of 50 data and each instance of the factor (of which there are five) takes one degree of freedom. The value of $\sqrt{2}t_{0.05'',45}s_r/\sqrt{10}$ is the LSD, that is, the maximum difference between means that we would accept as being not significant. If the means of each column are arranged in increasing order of magnitude then any difference between successive means greater than the LSD implies a significant difference.

4.6 ANOVA in Excel

Excel offers three "flavors" of ANOVA via its Analysis ToolPak: ANOVA: Single Factor; ANOVA: Two Factor with Replication; and ANOVA: Two Factor without Replication. The first is what we have just seen and accepts a data matrix set out as described, with variables in columns and repeats in rows. In this case the Grouped By: Columns radio button is checked. Column headers in the first row can be included, which helps with interpreting the output. A value of α, the probability at which the null hypothesis will be rejected, must be specified for an *F*-test, with 0.05 being the default. Thus the *F*-value is tested at the 95% probability level. The output looks like that in table 4.3, except that the terms "within groups" and "between groups" are used, and there are two extra columns. One has the probability associated with the calculated F-value and the last has the critical F-value for the designated α.

Example 4.1

Problem

Example 3.5 looked at comparing two means for the glucose levels in soft drinks being analyzed by a spectroscopic enzyme assay and an alternative enzyme electrode method. The conclusion of the analyses, using a t-test, was that the two means were different. The analytical laboratory therefore decided to check each method relative to an AOAC (Association of Official Analytical Chemists) method that employed HPLC. The analytical results for six replicate measurements (units mM) using each method are:

- Using the spectroscopic assay: 1.90, 1.82, 1.70, 1.94, 1.85, 1.90
- Using the enzyme electrode: 1.35, 1.65, 1.76, 1.41, 1.80, 1.33
- Using the AOAC method: 1.92, 1.82, 1.85, 1.79, 1.89, 1.95

Compare the means to determine whether there are significant differences between the methods at 95% probability.

Solution

An efficient way of determining whether there is a significant difference is to do a one-factor ANOVA. The means and 95% confidence intervals are calculated (see chapter 2) and plotted in figure 4.2.

ANOVA Example by Hand

1. First calculate the grand mean which is the mean of the all the data points. Therefore in Excel the overall mean is =AVERAGE(*range*) which for the above data included in spreadsheet 4.1 is =AVERAGE(A2:C7) giving the value 1.757.

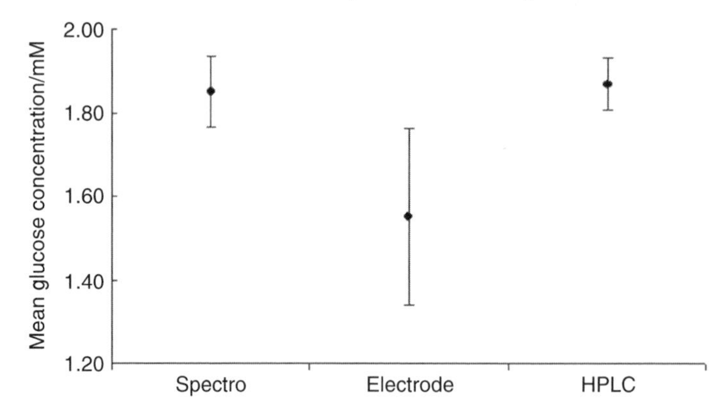

Figure 2 Means for the data in example 4.1. Error bars are ± one standard deviation.

2. Subtract the overall mean from each individual value and set up these values in a new array. That is, for the value in cell A2 which is placed in A17 input = (A2 – $B$12) which equals 0.014. Note $B$12 is the cell containing the grand mean.
3. Next calculate the between factors sum of squares, the within factor sum of squares, and the total sum of squares using the formulae given above.
4. Calculate the degrees of freedom. For between groups the degrees of freedoms are given by the number of instances of the factor (i.e., the number of columns) minus one, so for this example $3 - 1 = 2$. For within groups the degrees of freedom is the total number of measurements minus the number of instances of the factor $= N - k$, so for the present example this is $18 - 3 = 15$. Finally the total degrees of freedom is the total number of data points minus one, so $3 \times 6 - 1 = 17$.
5. Calculate the mean squares for the between and within factors which in each case is simply the sum of squares (SS_c and SS_r) divided by the degrees of freedom.
6. Divide the between factors mean square by the within factor mean squares to calculate the F-value. Recall the F-value is obtained by dividing the larger variance by the smaller variance. In the case of ANOVA the larger variance should be the between mean square and the between mean square is divided by the within mean square.
7. Finally from the F-value determine the associated probability using =FDIST(*F*, df_1, df_2) where again for ANOVA this

Spreadsheet 4.1

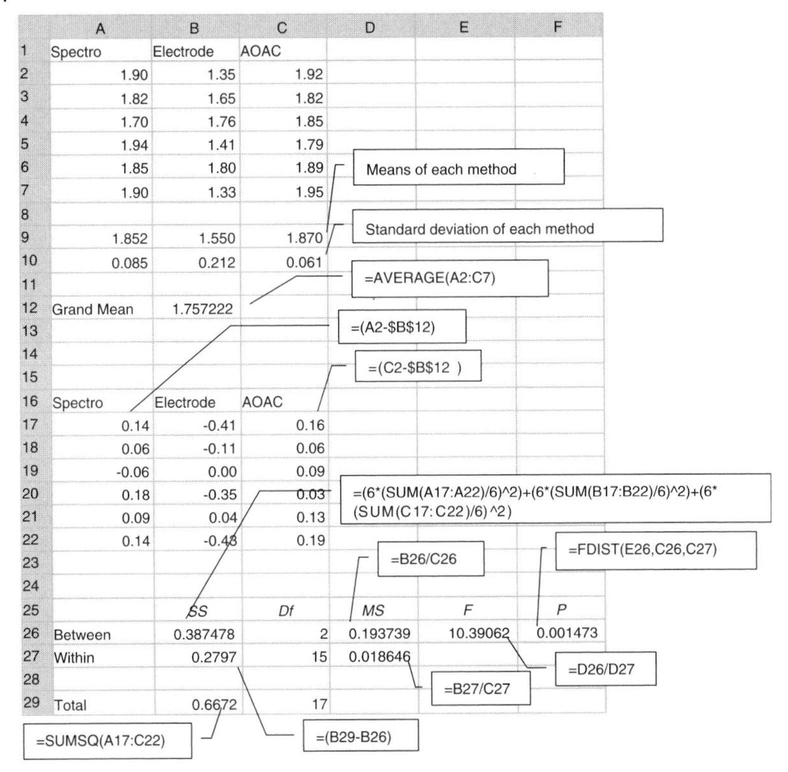

	A	B	C	D	E	F
1	Spectro	Electrode	AOAC			
2	1.90	1.35	1.92			
3	1.82	1.65	1.82			
4	1.70	1.76	1.85			
5	1.94	1.41	1.79			
6	1.85	1.80	1.89			
7	1.90	1.33	1.95			
8						
9	1.852	1.550	1.870			
10	0.085	0.212	0.061			
11						
12	Grand Mean	1.757222				
13						
14						
15						
16	Spectro	Electrode	AOAC			
17	0.14	-0.41	0.16			
18	0.06	-0.11	0.06			
19	-0.06	0.00	0.09			
20	0.18	-0.35	0.03			
21	0.09	0.04	0.13			
22	0.14	-0.48	0.19			
23						
24						
25		*SS*	*Df*	*MS*	*F*	*P*
26	Between	0.387478	2	0.193739	10.39062	0.001473
27	Within	0.2797	15	0.018646		
28						
29	Total	0.6672	17			

df_1 is the between factors degrees of freedom $(k-1)$ and df_2 is the within factors degrees of freedom $(N-k)$. If the value of the probability is $p < 0.05$ then we decide that there is a significant effect of the factor, which signifies that one at least one of the means is significantly different from the others.

Spreadsheet 4.1 shows the data with the ANOVA table of results at the bottom.

ANOVA Using Excel

To perform ANOVA calculations using Excel you need the Data Analysis plug in.

1. Arrange the data with the repeats for a given variable going down the columns and the different variables across the rows as in spreadsheet 4.1.

Spreadsheet 4.2

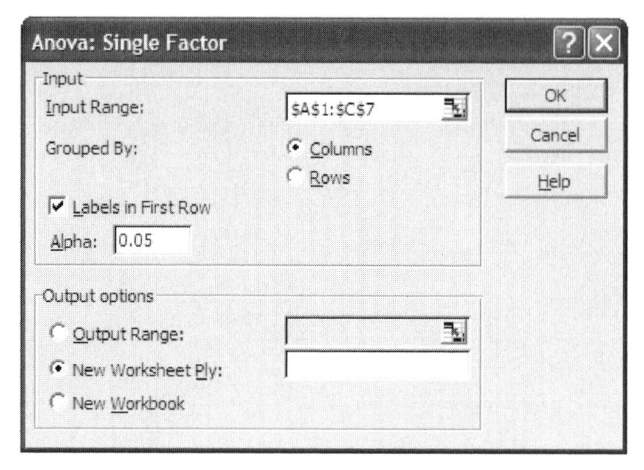

2. Go to the Tools menu and select Data Analysis. The Data Analysis dialog box will pop up.
3. From the Data Analysis dialog box select ANOVA: Single Factor. Single factor is chosen as only one factor is being varied, the method of analysis. A box will pop up which looks like spreadsheet 4.2.
4. Choose the range containing the data to be analyzed making sure the data are grouped by columns, that is, replicates going down a column and factors across the rows. Hence in spreadsheet 4.2 the range to be selected would be A1:C7.
5. Note two things. First, the data are grouped by columns, so columns radio button must be selected. Second, the data labels are included in the first row so the check box for "Labels in the first row" must be checked. Including the labels in the first row makes the interpretation of the results simpler, as Excel includes those labels in the output.
6. Select a probability, α, which in this case is 0.05 as we are assessing the significance at 95% probability.
7. Select a cell to be the top left hand corner of the output range. You need to ensure the output does not overlap with any data you may have on the spreadsheet. Excel does prompt you if there is danger of cells being overwritten. Alternatively you can select the output to be another New Worksheet or New Workbook.
8. The output looks like spreadsheet 4.3. The first table is a summary of the input data and the second table is the ANOVA

Spreadsheet 4.3

Anova: Single Factor						
SUMMARY						
Groups	*Count*	*Sum*	*Average*	*Variance*		
Spectro	6	11.11	1.851667	0.007297		
Electrode	6	9.3	1.55	0.04492		
AOAC	6	11.22	1.87	0.00372		
ANOVA						
Source of Variation	*SS*	*df*	*MS*	*F*	*P-value*	*F crit*
Between Groups	0.387478	2	0.193739	10.39062	0.001473	3.682317
Within Groups	0.279683	15	0.018646			
Total	0.667161	17				

output, which is the same as the table in spreadsheet 4.1 in which the ANOVA was calculated by hand. Note the output also includes the critical value of F, F_{crit}, for testing the calculated F at $\alpha = 0.05$.

9. The F-value is 10.39062 with an associated probability of 0.001473. As this is less than 0.05 it tells us the null hypothesis may be rejected at the 95% level (99.85% = 100 (1 – 0.001473) actually) and therefore we conclude that there is a significant difference in means due to the factor studied, that is, the different analytical methods. Hence, regardless of whether the ANOVA calculation was performed by hand or using Excel (thankfully) the same answer is arrived at.

Answer

The value of the probability is $p = 0.001473$ and therefore as $p < 0.05$ the null hypothesis is rejected with 95% probability and there is a significant effect due to the method of measuring the concentration of glucose in soft drinks.

Comments

1. Note that this is a fixed effect ANOVA, as the instances of the factor are confined to specific values, that is, the method of analysis is being chosen by the analyst.

2. The answer is that there is a significant effect but we do not know which of the parameters is responsible for this significant effect. One mean could be different from the others, but the result could imply all the means are different from each other. To solve this problem requires doing a LSD test. The LSD is calculated using

$$\mathrm{LSD} = \sqrt{2}\frac{t_{0.05'',N-k}s_{\mathrm{r}}}{\sqrt{n}}$$

where $s_{\mathrm{r}} = \overline{SS_r}$, n is the number of replicates (6), k is the number of instances of the factor, and $N-k$ is the degrees of freedom between factors (15). Therefore, for the data above

$$\mathrm{LSD} = \sqrt{2}\frac{t_{0.05'',15}0.018646}{\sqrt{6}} = 0.0229451$$

3. Using an Excel spreadsheet, the calculation of LSD would be written as = TINV(0.05, *df*)*s*SQRT(2)/SQRT(*n*) which for the above data would be = TINV(0.05, 15)*D27*SQRT(2)/SQRT(6). To use this number to determine which of the instances of the factor are significantly different we arrange the means in order of magnitude and determine the differences between successive means, that is, AOAC (1.870 mM), Spectro (1.852 mM), Electrode (1.550 mM), and therefore AOAC – Spectro = 0.018 mM, Spectro – Electrode = 0.302 mM. The difference in means between the spectroscopic method and the enzyme electrode method is considerably greater than LSD and the difference between the AOAC and the spectroscopic methods is just less than LSD. Therefore the enzyme electrode method differs significantly from the other two (you already knew this from example 3.5). Looking at figure 4.2 the conclusion appears quite reasonable.

4.7 Sampling

In analytical chemistry the test material analyzed is usually only a part of the system for which information is required. It is common for the system to be just too big to analyze all of it; for example, an ocean or river, or, as in the case of the chemical industry, there would be nothing left to sell if the total output of a synthesis were destructively analyzed. The chemist, therefore, takes only a small part as a representative sample, and assumes that the results of the analysis can be taken as the answer for the whole. ANOVA can be used to check this assumption, and to determine the variation in the test portions chosen and the contribution to the variation of the measurement process.

The total analysis variance is given by

$$\sigma^2_{\text{analysis}} = \sigma^2_{\text{measure}} + \sigma^2_{\text{sampling}}$$

where $\sigma^2_{\text{measure}}$ is the variance in making the measurement and $\sigma^2_{\text{sampling}}$ is due to actual differences between test portions.

Example 4.2

Problem

A grain silo is sampled at the top, middle, and bottom with four separate grab samples being taken at each level. The amount of protein in the grain of each sample is then determined by a Kjeldahl nitrogen analysis. The results are given in table 4.4. Does the sampling procedure have a significant effect on the results at the 95% probability level? If so, what are the standard deviation in sampling and the standard deviation in the analytical method? Therefore, determine the standard deviation expected of single measurements taken at random from anywhere in the silo.

Solution

This is treated as a one-way ANOVA with the factor studied being sampling position. Figure 4.3 shows the means and 95%

Table 4.4 Analysis of grain taken from different levels in a grain silo

Top (% protein)	*Middle (% protein)*	*Bottom (% protein)*
12.3	13.4	13.2
12.7	12.8	13.5
11.8	13.6	13.1
12.2	13.0	12.9

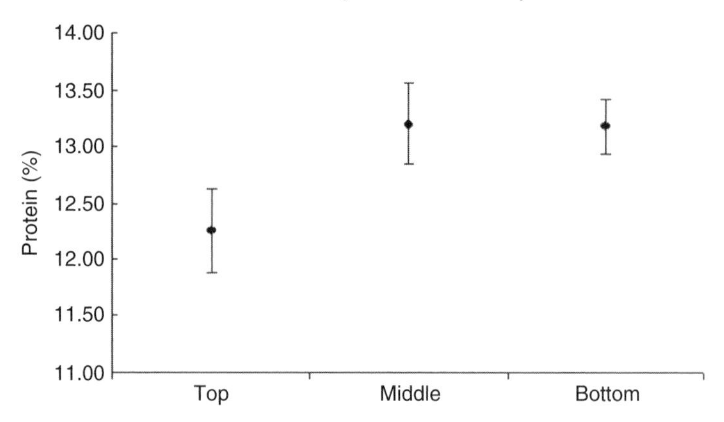

Figure 4.3 Means for the data in example 4.2. Error bars are ± one standard deviation.

confidence intervals of the results from each location.

1. An Excel spreadsheet is set up with each sampling location across the columns and the repeats for a given location down the rows.
2. Select ANOVA: One Factor from the Data Analysis dialog box add-in under the Tools menu. Choose the range of data to be analyzed, make sure the data are grouped by columns, select the probability (in this case $p = 0.05$ as we want to know whether the sampling variance is significant at a 95% confidence level), and select the output range.
3. The output ANOVA table is shown in spreadsheet 4.4.

As the probability associated with the F-value less than 0.05 ($0.004333 < 0.05$), or $F > F_{crit}$ ($10.57895 > 4.256492$), there is a significant effect due to sampling at the 95% probability level. The within groups mean square allows estimation of the repeatability of the measurement and so

$$s_r = \sqrt{(0.1108)} = 0.33\% \text{ protein}$$

Spreadsheet 4.4

Source of Variation	*SS*	*df*	*MS*	*F*	*P-value*	*F crit*
Between Groups	2.345	2	1.1725	10.57895	0.004333	4.256492
Within Groups	0.9975	9	0.110833			
Total	3.3425	11				

The parameter s_r is an estimate of $\sigma_{measure}$ the standard deviation of the measurement.

As we have seen, the between groups mean square ($\overline{SS}_c$) estimates the combination of the measurement variance (σ_r^2) and the variance due to the different sampling positions (σ_c^2), and from equation 4.3

$$\sigma_c \approx \sqrt{\frac{\overline{SS}_c - \overline{SS}_r}{n}} = \sqrt{\frac{1.1725 - 0.1108}{4}} = 0.52\% \text{ protein}$$

Hence the variance of a single analysis is given by

$$\sigma^2_{analysis} = \sigma^2_{measure} + \sigma^2_{sampling} = (0.33)^2 + (0.52)^2 = 0.376$$

and the standard deviation $\sigma_{analysis} = 0.61\%$ protein.

Answer

The variance in the sampling does have a significant effect on the overall measurement variance at the 95% level as determined by ANOVA. The standard deviation of sampling is 0.52% and the standard deviation of the analysis is 0.33% protein. The estimated standard deviation of single random samples is 0.61% protein.

Comment

It is common to find that the variability of sampling is greater than that of the analytical measurement. This is particularly so in environmental monitoring where the samples often show great variation. In choosing an analytical technique, there is no point in spending more funds in improving the precision of the method if the majority of the variance arises from sampling. In the example above, if the standard deviation of the analysis were reduced 10-fold (to 0.033% protein) the standard deviation of a single measurement would still be 0.52% protein.

4.8 Multiway ANOVA

The principles of ANOVA can be extended to two or more factors and the variances of each factor and their interactions can be

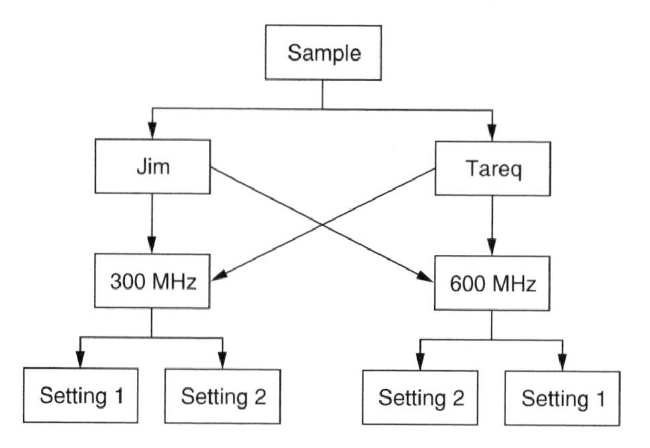

Figure 4.4 Experiments performed to investigate the effect of factors on the quantitative NMR of Dalapon sodium.

calculated and evaluated for significance. If measurements are made at every combination of each factor then the system is said to be cross-classified. If some of the factors are varied separately for each level of another factor then the system is nested. An example is a recent study of the effects of changing analyst, spectrometer, and instrument settings on the quantitative NMR analysis of the agricultural chemical Dalapon sodium. The experiments are laid out in figure 4.4. Each analyst analyses every sample on both machines and so the effects of analysts and machines are crossed. However, the settings on the 300 MHz machine are specific to it and cannot be used on the 600 MHz machine and vice versa. Hence the machine setting is a factor that is nested.

4.9 Two-Way ANOVA in Excel

Excel only caters for one- and two-way ANOVA. In two-way ANOVA there are two factors being considered. For example, we may be interested in the effect of changing the catalyst and the temperature in a synthesis. One factor is the catalyst (e.g., Zn or Li) and the other is temperature (e.g., 50, 70, or 90°C). In two-way ANOVA in Excel the distinction is made between measurements that are repeated and those for which only a single measurement is made, at each combination of factors. The layout for this example is given in table 4.5.

Table 4.5 Data grid for a two-way ANOVA: with replication (two measurements at each combination of factors) (right); and without replication (left)

Factor 2 = temperature	*Factor 1 = catalyst* *Zn*	*Li*
50°C		
70°C		
90°C		

Factor 2 = temperature	*Factor 1 = catalyst* *Zn*	*Li*
50°C		
50°C		
70°C		
70°C		
90°C		
90°C		

In the second case the data are duplicated. Replication of data allows independent estimation of the measurement (within variable) standard deviation in addition to estimation of the effect of the interaction of the factors. *What is an interaction?* If the effect of each factor is independent, the interaction is zero and the changes in the value of the measurand as a result of changing each factor add up to give the overall change observed. The result at a level of one factor is not at all influenced by the level of the other, and vice versa. However, for many systems the level of one factor does have an effect on the effect of the level of the other, and this is determined in the ANOVA. An example is the effects of time and temperature on the rate of a chemical reaction. At small times after the commencement of the reaction, increasing the temperature will speed up the reaction considerably, but as the reaction nears equilibrium at longer times the effect of the temperature is less. The result is an extra line in the ANOVA output table that quantifies this interaction effect in addition to the so-called main effects of the factors themselves. The input required is similar to the one-way ANOVA. For two-way with replication, with the data laid out as shown in table 4.5, the entire matrix of headers and data is selected as the input range. Note that there is no box to check for labels, as these must be included. It is necessary to specify how many repeats there are per combination of factor levels, and these must be the same. In the example of the Zn and Li catalysts the measurements are duplicated.

Spreadsheet 4.5

ANOVA

Source of Variation	SS	df	MS	F	P-value	F crit
Sample						
Columns						
Interaction						
Within						
Total						

The Excel output from the two-way ANOVA with replication has summary statistics of each factor with an ANOVA table that looks like spreadsheet 4.5.

Unfortunately, the labels are used for the statistics but do not find their way to the ANOVA table. The word "Sample" refers to the factor in the rows (having levels 50°C, 70°C, and 90°C in the example), the columns are in "Columns" (Zn and Li), "Interaction" is as described above, and "Within" is the residual or within factor variance. The interpretation of the columns of the ANOVA table is exactly as described above for the one-way case.

Example 4.3

Problem

Two students calibrated an autopipette, of volume 500 μL, using forward and reverse pipetting. Forward pipetting is where the pipette plunger is depressed to the first stop, an aliquot drawn into the pipette and the aliquot dispensed by depressing the pipette to the second stop where all the liquid is blown out of the pipette. In reverse pipetting the pipette is depressed to the second stop, the liquid drawn up, and dispensed by depressing the pipette to the first stop, thus leaving some liquid in the pipette. The amount of fluid dispensed by the pipette is determined by weighing. The data obtained by the two students Quinn and Martin with 10 replicates with each method are shown in spreadsheet 4.6 (where the masses shown have units of mg).

Determine whether there is any difference in the means with 95% probability with regards to either the two analysts or the two pipetting methods.

Spreadsheet 4.6

Quinn		Martin	
Forward	Reverse	Forward	Reverse
495.3	488.9	495.3	488.9
496.0	488.0	496	489.6
498.7	489.2	498.7	488
497.7	485.4	497.7	489.2
498.3	488.4	498.4	488.4
498.0	488.1	498	488.4
498.2	487.5	498.2	485.4
497.6	489.5	497.6	488.1
498.1	490.3	498.1	487.5
497.2	488.8	497.2	489.5

Solution

The solution to this problem involves performing a two-factor ANOVA with replication. This will be demonstrated using Excel. First we plot the means and standard deviations to give a first impression of the data. With two factors there are two options for our plot. We can have the analyst as the x-variable and plot the forward and reverse data on the graph, or vice versa with the direction as the x-variable and the analysts on the graph. This is shown in figure 4.5. To perform the calculation:

1. Arrange the data as shown in spreadsheet 4.8 with one factor going across the columns (the student in this case) and the other factor that contains replicates going down the column (forward or reverse pipetting).
2. Go to the Tools menu and select Data Analysis. The Data Analysis dialog box will pop up.
3. From the Data Analysis select ANOVA: Two Factors With Replication. A box will pop up which looks like spreadsheet 4.7.
4. Choose the range containing the data to be analyzed making sure the data are grouped with one factor and replicates down a column and the other factor across the rows. Hence, in spreadsheet 4.7 the range to be selected would be A1:C21.
5. Select the number of rows per sample. In this case there are 10 replicates for each method of pipetting by each student so the number of rows per sample is 10.
6. Choose the probability. In this example we want 95% confidence levels so for "Alpha" we insert 0.05.

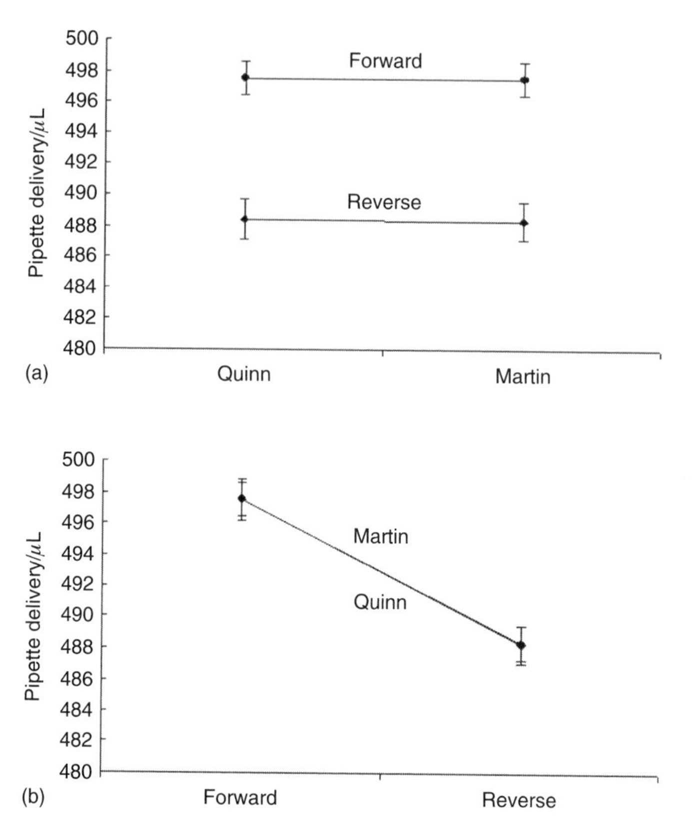

Figure 4.5 (a) Means of Quinn's forward and reverse results, and Martin's forward and reverse results. (b) Means of the forward pipetting results for Quinn and Martin, and reverse results for Quinn and Martin. Error bars are ± one standard deviation.

Spreadsheet 4.7

Anova: Two-Factor With Replication

Input

Input Range: $N$23

Rows per sample:

Alpha: 0.05

Output options

Output Range:

New Worksheet Ply:

New Workbook

OK

Cancel

Help

7. Select an output range (as before, the top left-hand cell) or you can select the output to be a New Worksheet or New Workbook.
8. The output looks like spreadsheet 4.8.
9. As can be seen from the output, there is a summary of each of the data for each factor followed by the ANOVA results table. Recall there are three key F-values calculated.

   (a) First, there is the row of data titled "Sample," which refers to a comparison between the sets of data down the rows. In this example the two sets of data in each column is a comparison between forward and reverse pipetting. As the probability associated with the F-value is 4.95×10^{-24}, which is <0.05, then we can say the null hypothesis is rejected and there is a significant difference between forward and reverse pipetting.
   (b) The second F-value is titled "Columns" and is a comparison between the factor across the columns. In the example this is a comparison between the two analysts. We can see that for this example $P = 0.8505$ which is >0.05 and therefore the null hypothesis is accepted: there is no difference in means between the two analysts Quinn and Martin.
   (c) The third row is titled "Interaction" and assesses whether the difference between means for one factor is influenced by the other. In terms of this example the F-value in this row is a test of whether the significant difference in means for forward versus reverse pipetting is dependent on the analyst or not. In this case the null hypothesis is that there is no interaction. In other words, the difference in mean is not dependent on the analyst. As can be seen from the output data, the probability associated with the F-value of 0.021723 is 0.88 which is >0.05 and therefore the null hypothesis is accepted: there is no relationship between the analyst and the difference in mean volume pipetted with the two pipetting techniques.

Spreadsheet 4.8

	A	B	C	D	E	F	G
1		Quinn	Martin				
2	Forward	495.3	495.3				
3	Forward	496.0	496				
4	Forward	498.7	498.7				
5	Forward	497.7	497.7				
6	Forward	498.3	498.4				
7	Forward	498.0	498				
8	Forward	498.2	498.2				
9	Forward	497.6	497.6				
10	Forward	498.1	498.1				
11	Forward	497.2	497.2				
12	Reverse	488.9	488.9				
13	Reverse	488.0	489.6				
14	Reverse	489.2	488				
15	Reverse	485.4	489.2				
16	Reverse	488.4	488.4				
17	Reverse	488.1	488.4				
18	Reverse	487.5	485.4				
19	Reverse	489.5	488.1				
20	Reverse	490.3	487.5				
21	Reverse	488.8	489.5				
22							
23	Anova: Two-Factor With Replication						
24							
25	SUMMARY	Quinn	Martin	Total			
26	*Forward*						
27	Count	10	10	20			
28	Sum	4975.359	4975.2	9950.559			
29	Average	497.5359	497.52	497.5279			
30	Variance	1.159004	1.175111	1.1057			
31							
32	*Reverse*						
33	Count	10	10	20			
34	Sum	4884.263	4883	9767.263			
35	Average	488.4263	488.3	488.3631			
36	Variance	1.773636	1.5	1.554865			
37							
38	*Total*						
39	Count	20	20				
40	Sum	9859.622	9858.2				
41	Average	492.9811	492.91				
42	Variance	23.22739	23.63779				
43							
44							
45	ANOVA						
46	*Source of Variation*	*SS*	*df*	*MS*	*F*	*P-value*	*F crit*
47	Sample	839.9382	1	839.9382	599.1265	4.95E-24	4.113161
48	Columns	0.050525	1	0.050525	0.036039	0.8505	4.113161
49	Interaction	0.030454	1	0.030454	0.021723	0.883648	4.113161
50	Within	50.46977	36	1.401938			
51							
52	Total	890.489	39				

Answer

The two-factor ANOVA shows there is a significant difference in the volume dispensed by a 500 μL autopipette if forward or reverse pipetting is employed, which is independent of whether Quinn or Martin used the pipette. There is no interaction between the effects.

Comment

The row that assesses the interaction gives important information. It was quite obvious from the data that the significant difference between the means for the different pipetting techniques was independent of the analyst. However, if the data were as in spreadsheet 4.9, then you would suspect that a significant difference in means between the two pipetting methods would be dependent on whether Quinn was the analyst. The output of the Two Factor with Replicates ANOVA for this data is shown in spreadsheet 4.10.
Now there is a significant difference between reverse and forward pipetting (see the *P*-value for the row labeled "Sample") and between Quinn and Martin (see the *P*-value for the row labeled "Columns"). The "Interaction" row also shows a significant *F*-value. We can interpret this as that the difference in the pipetting methods depends on the person doing the pipetting. This can be

Spreadsheet 4.9

Quinn		Martin	
Forward	Reverse	Forward	Reverse
495.3236	488.9168	495.3	497.2
496.0243	488.0158	496	498.3
498.7272	489.2171	498.7	499
497.7261	485.4131	497.7	498.7
498.3267	488.4163	498.4	496
498.0264	488.1159	498	497.5
498.2266	487.5153	498.2	498.8
497.626	489.5174	497.6	498
498.1265	490.3183	498.1	498.1
497.2256	488.8167	497.2	499
497.5359	488.4263	497.52	498.06
1.076571	1.331779	1.084025	0.947746

Each cell in this row gives the mean of the data in the column above

Each cell in this row gives the standard deviation of the data in the column above

Spreadsheet 4.10

ANOVA

Source of Variation	*SS*	*df*	*MS*	*F*	*P-value*	*F crit*
Sample	183.5963	1	183.5963	146.7018	2.94E-14	4.113161
Columns	231.257	1	231.257	184.7848	9.53E-16	4.113161
Interaction	232.7883	1	232.7883	186.0084	8.63E-16	4.113161
Within	45.05377	36	1.251494			
Total	692.6955	39				

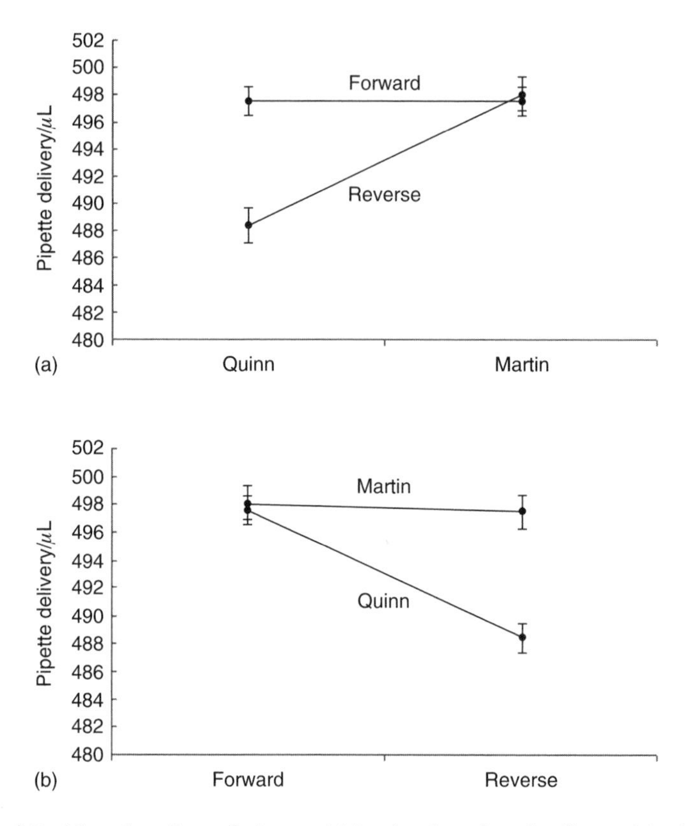

Figure 4.6 New data from Quinn and Martin plotted as for figure 4.5, showing an interaction effect.

seen in a plot of the means (figure 4.6). If the slopes of the two lines in either figure 4.6(a) or 4.6(b) are the same, then there is no interaction effect. If they are significantly different there is an effect. In figure 4.5 the lines in figure 4.5(a) are both nearly horizontal and in figure 4.5(b) overlaid. Although this is a special case where the effect of the analyst is zero, the slopes are the same and it is concluded there is no interaction effect.

4.10 Calculations of Multiway ANOVA

It is not possible to perform three-way or higher ANOVA in Excel, nor is it practical to perform the calculations manually. Many statistical packages do offer such analysis and require the data to be in a somewhat different form. The measurement results are in one column (sometimes known as the dependent variable) and each factor is represented by another column in which the level is given. Up to now we have considered situations in which the different levels of a factor are discrete entities—analysts, methods, etc. However, we have also referred to factors that are continuous variables, such as time and temperature. The model that ANOVA builds in each case is slightly different, and most software can cope with this. The output from different software programs varies but mostly contains the important information of the mean squares, F values and associated probabilities.

4.11 Variances in Multiway ANOVA

If the variance contributed by an effect must be known it can be calculated from the mean square of the factor. The within variables mean square is the square of the residual standard deviation as with two-way ANOVA (σ_r), that is, it is the variance left over after the variances due to all the effects have been extracted. For a system with no significant interaction effects, it is usually possible to treat the interaction variance as a residual variance, and in some software packages there is an option to rerun the analysis without interaction effects. In this case the variance of a factor is related to the mean square calculated in the ANOVA:

$$\overline{SS}_{\text{factor}} = \sigma_r^2 + n_{\text{factor}}\sigma_{\text{factor}}^2 \qquad (4.4)$$

therefore

$$\sigma_{\text{factor}} = \sqrt{\frac{\overline{SS}_{\text{factor}} - \overline{SS}_r}{n_{\text{factor}}}} \qquad (4.5)$$

where n_{factor} is the number of measurements made at each level of the factor.

5

Calibration

5.1 What This Chapter Should Teach You

- To describe the *linear calibration model* and how to estimate *uncertainties* in the calibration parameters and test concentrations determined from the model.
- To show how to perform calibration calculations using Excel.
- To calculate parameters and uncertainties in the *standard addition* method.
- To calculate *detection limits* from measurements of blanks and uncertainties of the calibration model.

5.2 Introduction

Calibration is at the heart of chemical analysis, and is the process by which the response of an instrument (in metrology called "indication of the measuring instrument") is related to the value of the measurand, in chemistry often the concentration of the analyte. Without proper calibration of instruments measurement results are not traceable, and not even correct. Scales in supermarkets are periodically calibrated to ensure they indicate the correct mass. Petrol pumps and gas and electric meters all must be calibrated and recalibrated at appropriate times.

A typical example in analytical chemistry is the calibration of a GC (gas chromatography) analysis. The heights of GC peaks are measured as a function of the concentration of the analyte in a series of standard solutions ("calibration solutions") and a linear equation fitted to the data. Before the advent of computers, a graph would be plotted by hand and used for calibration and subsequent measurement. Having drawn the best straight line through the points, the unknown test solution would be measured and the peak height read across to the calibration line then down on to the x-axis to give the concentration (figure 5.1). Nowadays, the regression equation is computed from the calibration data and then inverted to give the concentration of the test solution. Although the graph is no longer necessary to determine the parameters of the calibration equation, it is good practice to plot the graph as a rapid visual check for outliers or curvature. Because we can choose what values the calibration concentrations will take, the concentration is the *independent* variable, with the instrumental output being the *dependent* variable (because the output of the instrument depends on the concentration).

The example given in figure 5.1 is an example of a *univariate*, *linear* calibration. Univariate means that only one quantity is measured to establish the relationship (in the example the quantity is peak height). In modern instrumentation it is possible to collect many variables and use all the data to calibrate. For example, spectrometers offer digital output at hundreds or thousands of wavelengths, a mass spectrum can give peaks at thousands of mass-to-charge ratios, and inductively coupled plasma atomic emission spectrometry (ICPAES) gives the absorbances of many element emission transitions simultaneously.

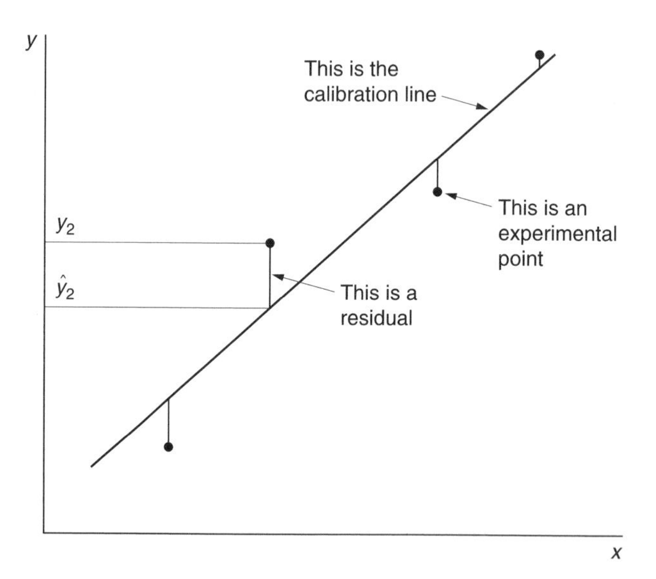

Figure 5.1 Schematic of calibration and measurement by a linear calibration equation.

All of these data are *multivariate* and mathematical analysis provides calibration models, pattern recognition, and other methods to extract useful information from this plethora of data. A linear calibration model is one that has linear coefficients in the quantities, but the concept is often used in the more restricted sense of linearity in the independent variable. There is no reason why relations used for calibration should in fact be linear in the independent variables. We know from experience that most relationships in chemistry are not strictly linear, but the conditions of use of an instrument or method have usually been arranged to be "in the linear range." There are advantages in having a linear measurement model: in the past it was by far the easiest to graph, it has a well-understood statistical model, and data can be fitted to an equation with unique parameters. However, spreadsheets have opened up the world of nonlinear fitting of data. "Solver" in Excel can perform a limited iterative search to minimize a function by changing specified coefficients, and there may be good reason to set up a nonlinear calibration rather than worry about whether or not the data are in some arbitrary linear range. However, a linear model is still used in the majority of chemical calibrations and therefore this chapter concentrates on linear calibration.

5.3 Linear Calibration Models

Before an analysis is performed using a particular method and instrument, there must be a measurement model, an equation that relates the quantities that are to be measured to the indication of the instrument and other influence quantities such as temperature and pressure. From this model we derive the calibration equation. Usually, the experiment is performed in such a way as to fix influence quantities and the calibration equation is determined in terms of the concentration of the measurand and some output of the instrument, which might be peak height, peak area, potential, current, absorbance, etc. A simple linear relation is

$$Y = a + bx \qquad (5.1)$$

where Y is the indication of the instrument (i.e., the instrument response), x is the independent variable, which in most cases for our purpose may be the concentration of the measurand, and a and b are the coefficients of the model and are known as the *intercept* and *slope* and are determined from a number of measurements of Y at particular values of x. The terms "intercept" and "slope" come from the days of graphs, a being the Y value when x is zero (i.e., the intercept on the Y axis) and b the rate of increase in Y with respect to x. Rather than be tied to the graphical descriptions, which are no longer necessary anyway, it is better to refer to b as the "analytical sensitivity." It has units of y/x. a is the expected indication of the instrument when x is zero, and therefore may be considered the "indication of the blank." The intercept has the same units as y. We decide on the values of x before the experiment (hence independent variable) and measure the indication of the instrument Y.

Any particular measurement of Y (y_i) will be subject to measurement error (ε_i), therefore

$$y_i = Y_i + \varepsilon_i \tag{5.2}$$

or

$$y_i = a + bx_i + \varepsilon_i \tag{5.3}$$

We shall see that one of our assumptions in making our calibration is that the uncertainty in the independent variable is much less that in the dependent variable, hence only one error term, ε_i in Y_i, is included. Be aware, however, that there is uncertainty in the dependent variable, and before using the model you must assure yourself that this uncertainty is sufficiently small to be neglected. The process of calibration involves the collection of data, namely values of y_i at a number of x_i, then fitting the model of equation 5.3 to the data.

Having established the values of a and b by calibration, measurement of the instrument response of an unknown material (y_0) allows calculation of its corresponding value of x, and uncertainty (see section 5.4.1):

$$\hat{x} = \frac{y_0 - a}{b} \tag{5.4}$$

As our calculation is an estimate of the quantity based on the instrumental response and calibration, the x gains a "hat". There is uncertainty in the value of $\hat{x}$ because of the measurement uncertainty in establishing a and b and additionally in the subsequent measurement of y_0. Because we know, or assume, the statistical properties of the calibration equation although we do not know the ε_i, it is possible to calculate confidence intervals on the parameters a and b and then on the estimated x.

Analytical methods for which a blank reading may be made and subtracted from subsequent results rely on this procedure to force the calibration through zero. The calibration model is now

$$Y = bx \tag{5.5}$$

and

$$y_i = bx_i + \varepsilon_i \tag{5.6}$$

and the estimate of the concentration is thus

$$\hat{x} = \frac{y_0}{b} \tag{5.7}$$

5.3.1 Determination of *a* and *b*

Even for a simple equation like equation 5.1, by making different assumptions about the data we will arrive at different values of a and b. Most spreadsheets and calculators perform "classic" linear regression. The assumptions are:

1. The linear model is correct (i.e., the response of the measuring instrument does indeed vary linearly with concentration).
2. All uncertainty resides in the dependent variable (Y) and is normally distributed.
3. The data are known as homoscedastic, which means that the errors in y are independent of the concentration. Data for which the uncertainty, for example, grows with the concentration are heteroscedastic data.

Most chemical systems break one or other of these assumptions, but for reasonably linear data it is not seen to be of too great a concern. It must be noted that, for some analyses, other types of regression should be contemplated. For example, in ultratrace analysis, the successive dilutions that must be done to achieve suitably low concentrations of calibration solutions lead to uncertainty in the independent variable that is not negligible. In addition, modern instrumentation and methods that deliver reduced measurement uncertainty (in the dependent variable) lead to a situation in which we can no longer ignore the contribution of this uncertainty. So-called "total" least squares, or the "errors in variables" model can be used in these cases.

Data are often heteroscedastic. A relative standard deviation of 2% that applies across a range of concentrations means that the standard deviation of a measurement of 1 mM is 0.02 mM, but is 2 mM when a measurement of 0.1 M is made. If this is the case the regression should be performed by weighting the data by $1/\sigma^2$, where σ is the standard deviation of the measurement. Thus results with smaller uncertainty must be fitted more closely than those with greater uncertainty (Miller and Miller—see Bibliography).

The assumptions given above lead to a desire to minimize $s_{y/x}$ which is known as the standard error of the regression:

$$s_{y/x} = \sqrt{\frac{\sum_{i=1}^{i=n} (y_i - \hat{y}_i)^2}{df}} \tag{5.8}$$

$s_{y/x}$ has the same units as y. y_i is a measured value of the dependent variable and $\hat{y}_i$ is the estimated value from the regression equation (equation 5.2 or equation 5.5). The sum is over the n calibration data. The difference $(y_i - \hat{y}_i)$ in equation 5.8 is known as the residual for the obvious reason that it is what remains of a measured value of Y (y_i) when the estimated value, $\hat{y}_i$, is subtracted from it. It is said, therefore, that regression minimizes the sum of the squares of the residuals. The degrees of freedom, df, is $n-2$ for the calibration model in equation 5.1 because two coefficients (a and b) are calculated in the model, and df is $n-1$ for the calibration model in equation 5.5, for which only the analytical sensitivity is calculated. If the data fit the equation perfectly then $y_i = \hat{y}_i$, all the residuals are zero, and so $s_{y/x} = 0$. Note that a point that has a large residual adds disproportionately to $s_{y/x}$ because of the square. This means that the regression will avoid one very large residual in favor of a number of smaller ones, and so a single rogue point can throw out the whole calibration. The tendency of a point to drag the line toward it is known as "leverage."

The coefficients of the equations are calculated from the data by

$$b = \frac{\sum_{i=1}^{i=n} [(x_i - \overline{x})(y_i - \overline{y})]}{\sum_{i=1}^{i=n} (x_i - \overline{x})^2} \tag{5.9}$$

$$a = \overline{y} - b\bar{x} \tag{5.10}$$

where a bar indicates the mean of all the x or all the y data. That is, $\overline{x}$ is the average concentration of all the calibrator concentrations used to establish the calibration curve and $\overline{y}$ is the average of all the measured responses (averaging all measured responses means summing the response for every individual calibrator and dividing by the number of calibrators).

Remember that the slope has units of the units of y over the units of x, and the intercept has the units of y. In undergraduate reports failure to include proper units invariably incurs the wrath of a demonstrator. In the real world it makes you look unprofessional, so try not to forget.

Standard deviations may be calculated for the coefficients:

$$s_b = \frac{s_{y/x}}{\sqrt{\sum_{i=1}^{i=n} (x_i - \overline{x})^2}} \tag{5.11}$$

$$s_a = s_{y/x}\sqrt{\frac{\sum_{i=1}^{i=n} x_i^2}{n\sum_{i=1}^{i=n} (x_i - \overline{x})^2}} \tag{5.12}$$

Confidence intervals on the slope and intercept are determined by multiplying the standard deviations by a two-tailed Student t-value (recall from chapter 2 that two tails refers to both halves of the distribution) at an appropriate probability and degrees of freedom of the regression. These can be used when the slope or intercept is needed for further calculations, for example in determining the activation energy from the slope of an Arrhenius plot ($\log(k)$ against $1/T$).

$$b \pm t_{\alpha'',df} s_b \tag{5.13}$$

$$a \pm t_{\alpha'',df} s_a \tag{5.14}$$

Recall here that $df = n - 2$, because two parameters are calculated from n data. Of interest to analytical chemists is the uncertainty imparted to the estimate of a concentration determined from the calibration equation. The standard deviation of the estimate of concentration from m measurements of an unknown sample giving mean response y_0 is

$$s_{\hat{x}_0} = \frac{s_{y/x}}{b}\sqrt{\frac{1}{m} + \frac{1}{n} + \frac{(y_0 - \overline{y})^2}{b^2 \sum_{i=1}^{i=n} (x_i - \overline{x})^2}} \quad (5.15)$$

This is an important equation as it allows the calculation of the information you are looking for, the standard deviation in the determination of the concentration of the unknown sample. All the other symbols refer to the calibration, including n, $\overline{x}$, and $\overline{y}$. Equation 5.15 is most instructive in telling us about what makes good calibrations and measurements. To make the standard deviation of the result as small as possible the standard error of the regression, $s_{y/x}$, must be small (i.e., the calibration fit should be good) and the analytical sensitivity (b) should be large (i.e., a small change in concentration should cause a large change in the instrument response). Looking at the terms under the square root sign, more calibration points (n) and more repeats of the unknown (m) are better, but there is a law of diminishing returns. Consider the $1/m$ and $1/n$ terms. If you only measure your unknown once then the term has the value 1. It is halved if the measurement is duplicated, but to halve it again four measurements must be done. The best number of points in the calibration curve is more tricky to decide, because n also appears in the degrees of freedom of $s_{y/x}$ and the Student t-value if a confidence interval is calculated. Six independently prepared solutions of different concentrations should be considered a minimum to establish a calibration equation, and ten is better. If the calibration curve is to be used for a number of determinations of unknowns it is certainly worth taking the time to get it right. The third term under the square root in equation 5.15 is zero if the measured response of the unknown happens to be at the mean of the calibration responses, that is, in the middle of the range ($y_0 - \overline{y} = 0$). As the unknown response moves toward the extremes of the calibration range this contribution to the uncertainty increases. Remember you should never use a calibration equation to estimate an unknown outside the range of concentrations that were used to establish the equation. Figure 5.2 shows the calibration line and 95% confidence interval for measured concentrations for different sets of data to illustrate the points made.

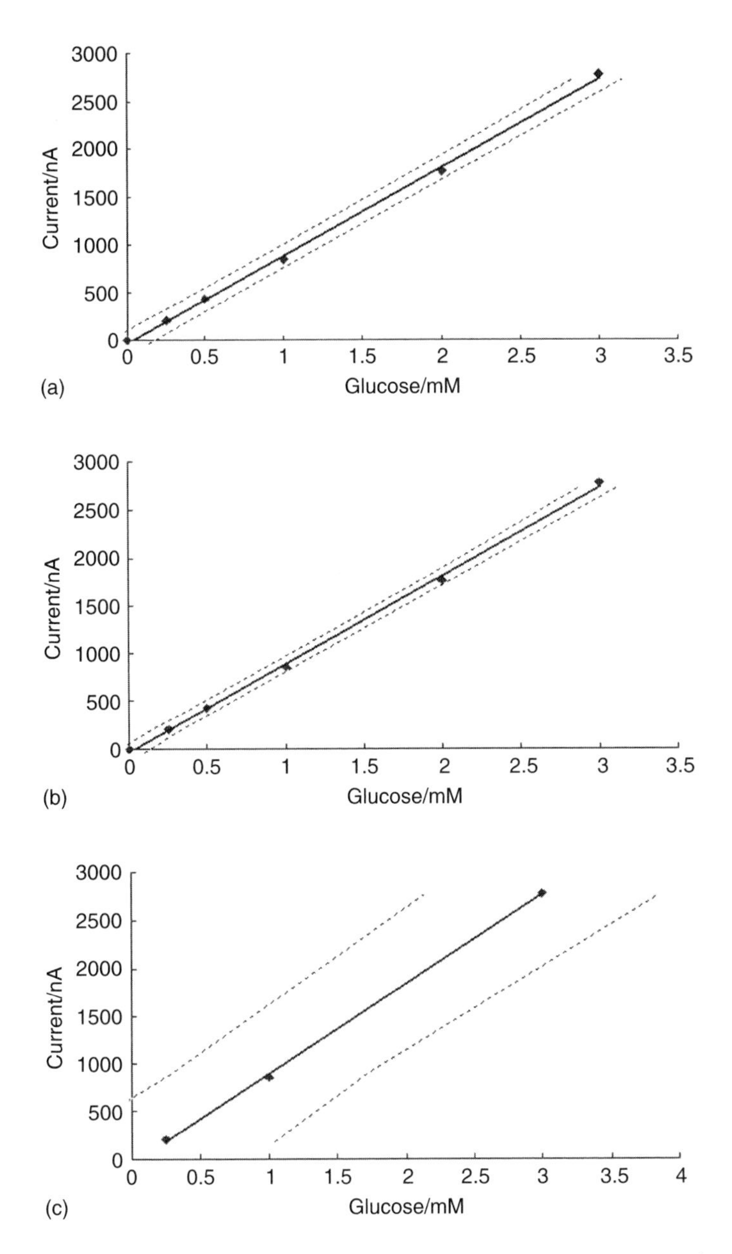

Figure 5.2 Calibration line and 95% confidence intervals on concentrations calculated from measurements of unknowns. Data are from calibration of a glucose monitor. (a) Six calibration solutions with a single measurement of the test solution; (b) six calibration solutions with three measurements of the test solution; (c) three calibration solutions with a single measurement of the test solution.

Table 5.1 Calibration of the analysis of glucose by a spectroscopic enzyme assay

[Glucose] (mM)	0.000	0.050	0.100	0.200	0.400	0.600	0.800
Absorbance	0.000	0.057	0.119	0.221	0.383	0.599	0.730

Example 5.1

Problem

To determine the concentration and associated uncertainty of glucose in a wine sample of unknown concentration using a spectroscopic enzyme assay where the calibration data is given in table 5.1.
Wine of unknown glucose concentration and calibration solutions of glucose were treated in the same way as follows: Wine (200 μL) or glucose solution (200 μL) was diluted to 5 mL in a volumetric flask by the addition of enzyme solution and buffer. Triplicate measurements of the absorbance of the treated wine solution were 0.253, 0.243, 0.238.

Solution

1. First it is always a good idea to plot the data as a visual aid. Looking at the plot in figure 5.3 a linear calibration model may be valid.
2. Calculate the coefficients of the linear calibration model a and b using equations 5.10 and 5.9. In equation 5.9, x_i and y_i are the individual values of x and y (i.e., the response actually measured at each concentration), $\overline{x}$ is the means of all the values of x_i used in the calibration, and $\overline{y}$ is the mean of all the measured responses. Therefore

$$\overline{x} = (0 + 0.05 + 0.1 + 0.2 + 0.4 + 0.6 + 0.8)/7$$

$$= 0.3071\,\text{mM}$$

$$\overline{y} = (0 + 0.057 + 0.119 + 0.221 + 0.383 + 0.599 + 0.730)/7$$

$$= 0.3013$$

Hence to calculate b use

$$b = \frac{\sum_{i=1}^{i=n}[(x_i - \overline{x})(y_i - \overline{y})]}{\sum_{i=1}^{i=n}(x_i - \overline{x})^2}$$

$$\frac{\{[(0 - 0.3071) + \cdots + (0.8 - 0.307)] \times [(0 - 0.301) + \cdots + (0.730 - 0.301)]\}}{[(0 - 0.3071) + \cdots + (0.8 - 0.307)]^2} = 0.9197\,\text{mM}^{-1}$$

and to calculate a use

$$a = \overline{y} - b\overline{x} = 0.3013 - (0.9197 \times 0.3071) = 0.0188$$

3. The next step is to calculate the standard error of the regression $s_{y/x}$. First, we must calculate $\hat{y}_i$ the value of Y estimated from the regression for each x_i. Hence for each concentration of the standards, x_i, $\hat{y}_i$ can be calculated using the regression equation we have now established: $\hat{y} = 0.0188 + 0.9197x_i$. The calculated values of $\hat{y}_i$ for each x_i are shown in spreadsheet 5.1. Now we can determine $s_{y/x}$ using

$$s_{y/x} = \sqrt{\frac{\sum_{i=1}^{i=n}(y_i - \hat{y}_i)^2}{df}}$$

$$= \sqrt{\frac{[(0 - 0.0188) + (0.057 - 0.0648) + \cdots + (0.730 - 0.7545)]^2}{5}}$$

$$= 0.0212$$

4. The next step is to calculate the standard deviations for the coefficients of the calibration a and b using equations (5.12) and (5.11):

$$s_b = \frac{s_{y/x}}{\sqrt{\sum_{i=1}^{i=n}(x_i - \overline{x})^2}} = \frac{0.0212}{\sqrt{[(0 - 0.3071) + \cdots + (0.8 - 0.307)]^2}}$$

$$= 0.0285\,\text{mM}^{-1}$$

$$s_a = s_{y/x}\sqrt{\frac{\sum_{i=1}^{i=n}x_i^2}{n\sum_{i=1}^{i=n}(x_i - \overline{x})^2}}$$

Spreadsheet 5.1

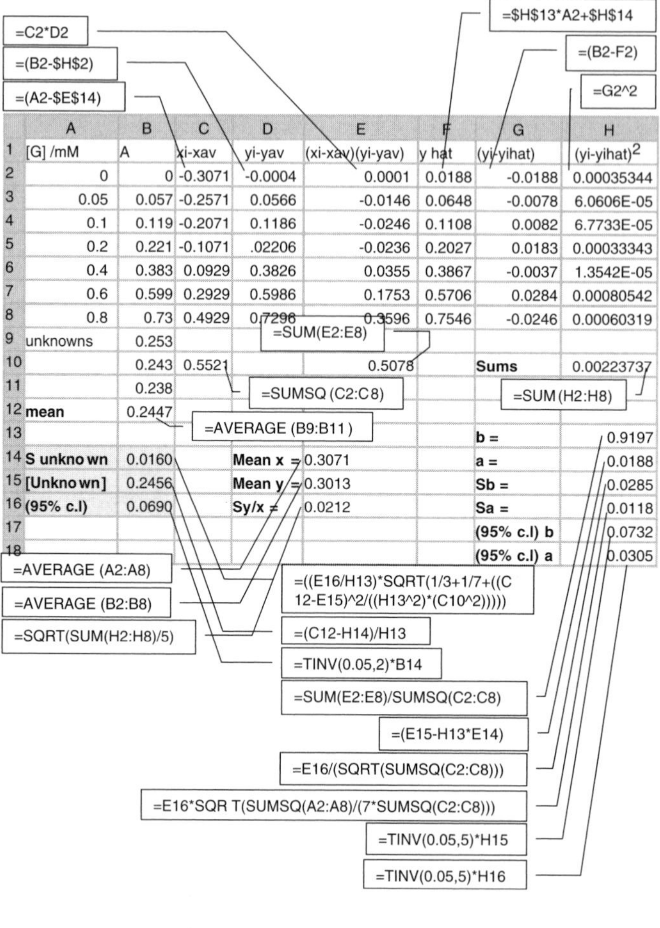

	A	B	C	D	E	F	G	H
1	[G] /mM	A	xi-xav	yi-yav	(xi-xav)(yi-yav)	y hat	(yi-yihat)	(yi-yihat)2
2	0	0	-0.3071	-0.0004	0.0001	0.0188	-0.0188	0.00035344
3	0.05	0.057	-0.2571	0.0566	-0.0146	0.0648	-0.0078	6.0606E-05
4	0.1	0.119	-0.2071	0.1186	-0.0246	0.1108	0.0082	6.7733E-05
5	0.2	0.221	-0.1071	.02206	-0.0236	0.2027	0.0183	0.00033343
6	0.4	0.383	0.0929	0.3826	0.0355	0.3867	-0.0037	1.3542E-05
7	0.6	0.599	0.2929	0.5986	0.1753	0.5706	0.0284	0.00080542
8	0.8	0.73	0.4929	0.7296	0.3596	0.7546	-0.0246	0.00060319
9	unknowns	0.253						
10		0.243	0.5521		0.5078		Sums	0.00223737
11		0.238						
12	**mean**	0.2447						
13							**b =**	0.9197
14	**S unknown**	0.0160		**Mean x =**	0.3071		**a =**	0.0188
15	**[Unknown]**	0.2456		**Mean y =**	0.3013		**Sb =**	0.0285
16	**(95% c.l)**	0.0690		**Sy/x =**	0.0212		**Sa =**	0.0118
17							**(95% c.l) b**	0.0732
18							**(95% c.l) a**	0.0305

$$= 0.02115 \times \sqrt{\frac{0^2 + 0.05^2 + \cdots + 0.8^2}{7 \times [(0 - 0.3071) + \cdots + (0.8 - 0.307)]^2}}$$

$$= 0.0118$$

where $n = 7$ is the number of data in the calibration. More informative are the confidence intervals on the slope and intercept. These are derived in the usual way:

$$b \pm t_{\alpha'', df} s_b = b \pm t_{0.05,5} s_b = 0.920 \pm 2.57 \times 0.0285$$

$$= 0.920 \pm 0.073\,\text{mM}^{-1}$$

$$a \pm t_{\alpha'', df} s_a = a \pm t_{0.05,5} s_a = 0.019 \pm 2.57 \times 0.0118$$

$$= 0.019 \pm 0.031$$

Recall that the t-value can be determined in Excel using =TINV(α, df), which here is =TINV(0.05, 5) = 2.57.

5. We now get to what we really wanted to know, how to calculate the uncertainty of the concentration of the test solution. This is done using equation 5.15. The triplicate measurements of the absorbances of the unknown are 0.253, 0.243, 0.238 which have a mean of $y_0 = 0.245$. Therefore the estimate of the concentration $\hat{x}$ is

$$\hat{x} = \frac{y_0 - a}{b} = \frac{0.02445 - 0.0188}{0.9197} = 0.246\,\text{mM}$$

and the uncertainty in $\hat{x}$ is

$$s_{\hat{x}_0} = \frac{s_{y/x}}{b}\sqrt{\frac{1}{m} + \frac{1}{n} + \frac{(y_0 - \bar{y})^2}{b^2 \sum_i (x_i - \bar{x})^2}}$$

$$= \frac{0.0212}{0.9197}\sqrt{\frac{1}{3} + \frac{1}{7} + \frac{(0.0245 - 0.3013)^2}{(0.9197)^2 \times [(0 - 0.3071) + \cdots + (0.8 - 0.307)]^2}}$$

$$= 0.0161\,\text{mM}$$

Therefore the 95% confidence interval about the estimate is

$$\pm t_{0.05,5} s_{\hat{x}_0} = \pm(2.57 \times 0.0161) = \pm 0.041\,\text{mM}$$

Now recall that this is the concentration of glucose in the test sample which contained only 200 μL of wine in a 5 mL flask. Therefore the concentration in the wine and its uncertainty is

$$(0.246 \pm 0.041) \times 5/0.2 = 6.2 \pm 1.0\,\text{mM}$$

Answer

The concentration of glucose in a wine sample and the associated uncertainty, quoted as a 95% confidence interval, is 6.2 ± 1.0 mM.

Comments

1. Plotting the data allows you to have a feel for whether the linear calibration model is valid or not, but be careful not to add a regression line until after you have had a careful look at the graph. Compare the two plots shown in figure 5.3. These plot the same data, but whereas figure 5.3(a) looks as if there may be a slight curve to the data, once the trend line is added in figure 5.3(b) it is much more difficult to ascertain whether or not the data are curved. This is not helped by the default line thickness of the Trendline feature in Excel being a thick 1.5 points.
2. An Excel spreadsheet, with some of the cell operations shown, which could be used to perform the above calculation

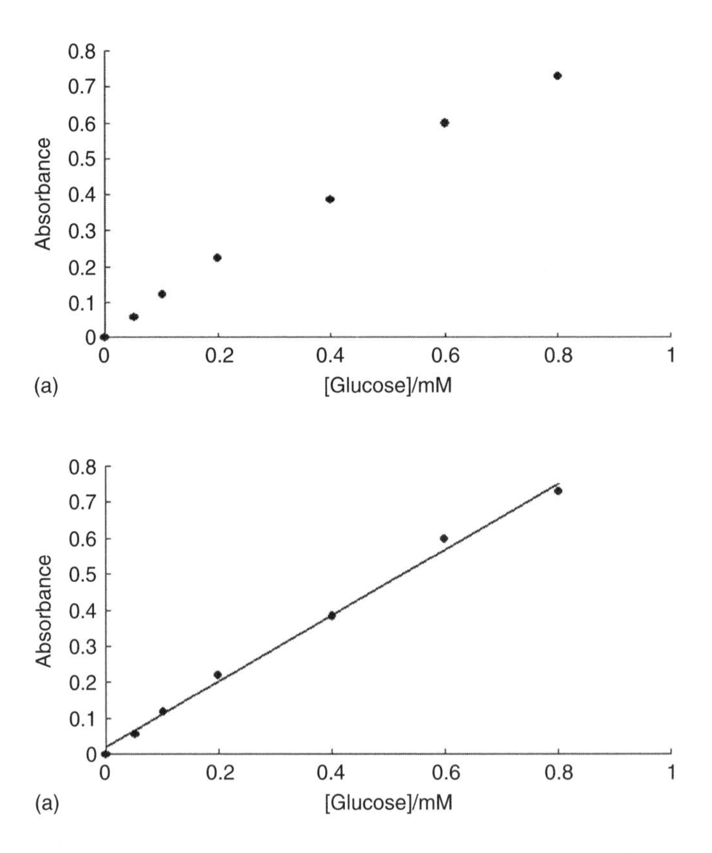

Figure 5.3 Calibration of a spectrophotometric analysis of glucose: (a) plot of the data; (b) with TRENDLINE.

is shown as spreadsheet 5.1. We shall explain how to simplify the calculations using Excel functions in a later section.

5.3.2 Outliers and the linear range

Two concerns of the analyst when calibrating a method are to ensure that each of the responses measured for the calibration solutions are suitably accurate, and that the range of concentrations chosen are in the range of validity of the calibration model (for a general discussion of how to choose a suitable calibration range see section 5.7). For a linear model, this is known as the linear range. A typical situation when there is a doubt about linearity is when the response of an instrument saturates at high concentration. The breakdown of the Beer–Lambert law of optical absorbance is an example of this (figure 5.4). One way of testing for the linear range is to fit the likely range from zero then predict the next point. The residual is then tested against $s_{y/x}$ of the linear fit:

$$t = \frac{\left|\hat{y} - y_{\mathrm{expt}}\right|}{s_{y/x}} \tag{5.16}$$

It is necessary to anchor the line at zero, so a regression with zero intercept is chosen. This restricts the method, but it is a common occurrence in chemistry when the data have been corrected for the blank. Figure 5.4 shows this procedure for data from the dye Sunset

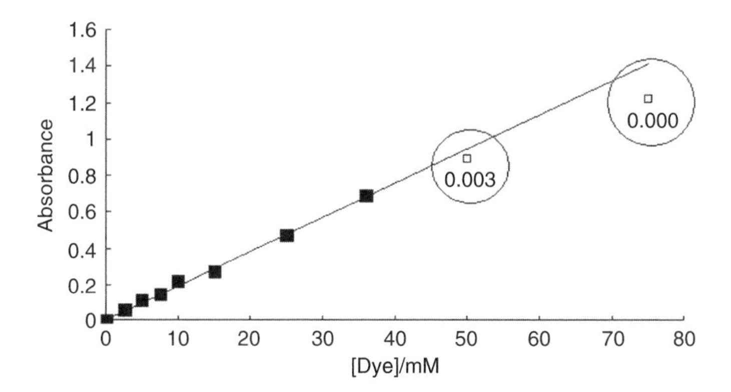

Figure 5.4 Calibration of a spectrometer with a dye showing saturation at the higher concentrations. The values on points are the probabilities of t-values determined by the procedure of predicting the next point (see text).

Yellow. By trial and error find the range for which all the subsequent points fail a significance test (i.e., the probability of $T > t$ is less than 0.05).

Unfortunately, sometimes the relation is nonlinear at all concentrations and when plotted the calibration line is a continuous curve. Beware of plotting a trend line through the data too soon as the straight line fools your eye into thinking the underlying data are also linear. It is always a good idea to plot a graph of the residuals against the concentrations. This magnifies any trends or discrepancies of the calibration line and allows a quick and usually accurate assessment of the quality of the fit. A well-behaved calibration has residuals randomly distributed about zero. Figure 5.5 illustrates different residual plots including cases where something has gone wrong. In the case of a suspect outlier the regression should be recalculated without the suspect point and the residual plot replotted including the suspect point, when the outlier should be even more pronounced. Statistical assessment of outliers is provided in a number of computer software applications, but none is easily implemented by hand or in a simple spreadsheet.

A rule of thumb is to suspect a point that has a residual of magnitude greater than $3s_{y/x}$. However, you should not have outliers in your calibration. If necessary solutions should be remeasured and if the response still appears anomalous, the solution should be made up again. You have control over your calibration, and if there is a problem with it, you must rectify the situation before any measurements of unknowns are made. Do not rely on statistics to bail you out of poor chemical technique!

Example 5.2

Problem

The mass fraction of calcium in a milk sample was analyzed using atomic absorption spectrometry (AAS). The calibration data are shown in table 5.2.

The calibration plot with linear regression is shown in figure 5.6. The regression equation is $y = 0.0177x + 0.1082$.

Determine whether a linear calibration model is suitable for these data.

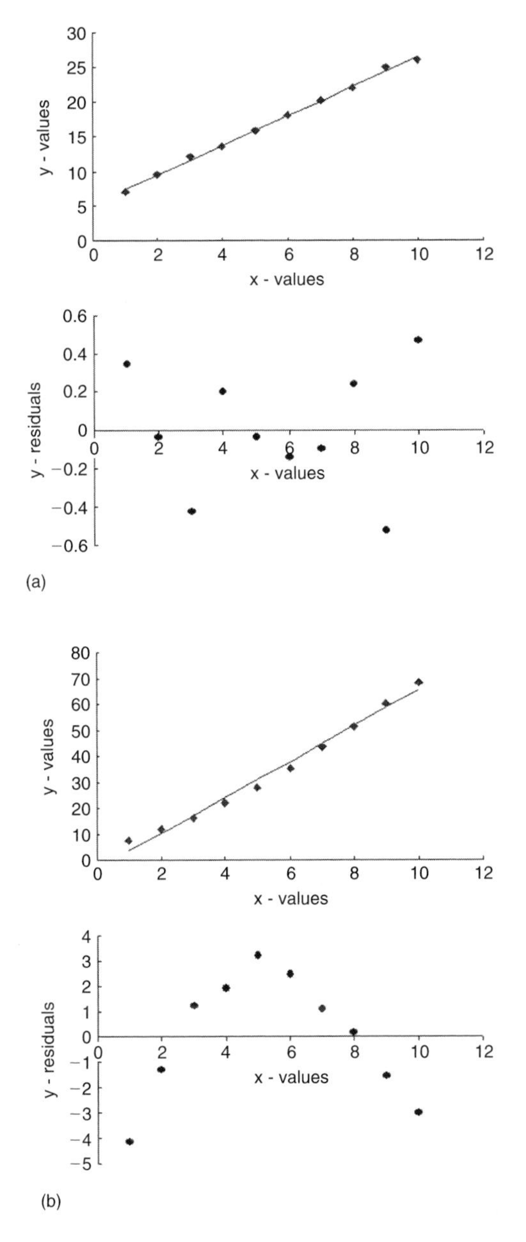

Figure 5.5 Calibration and residual plots illustrating different types of data. (a) Appropriate data for linear regression with normally distributed error in the y data only; (b) curvature throughout the range; (c) heteroscedacity with increasing standard deviation of responses with increasing concentration; (d) curvature at high concentrations—the dashed lines show the limiting linear and saturation values; (e) an outlier—the dashed line in the calibration plot and open circles in the residual plot are with the outlier removed.

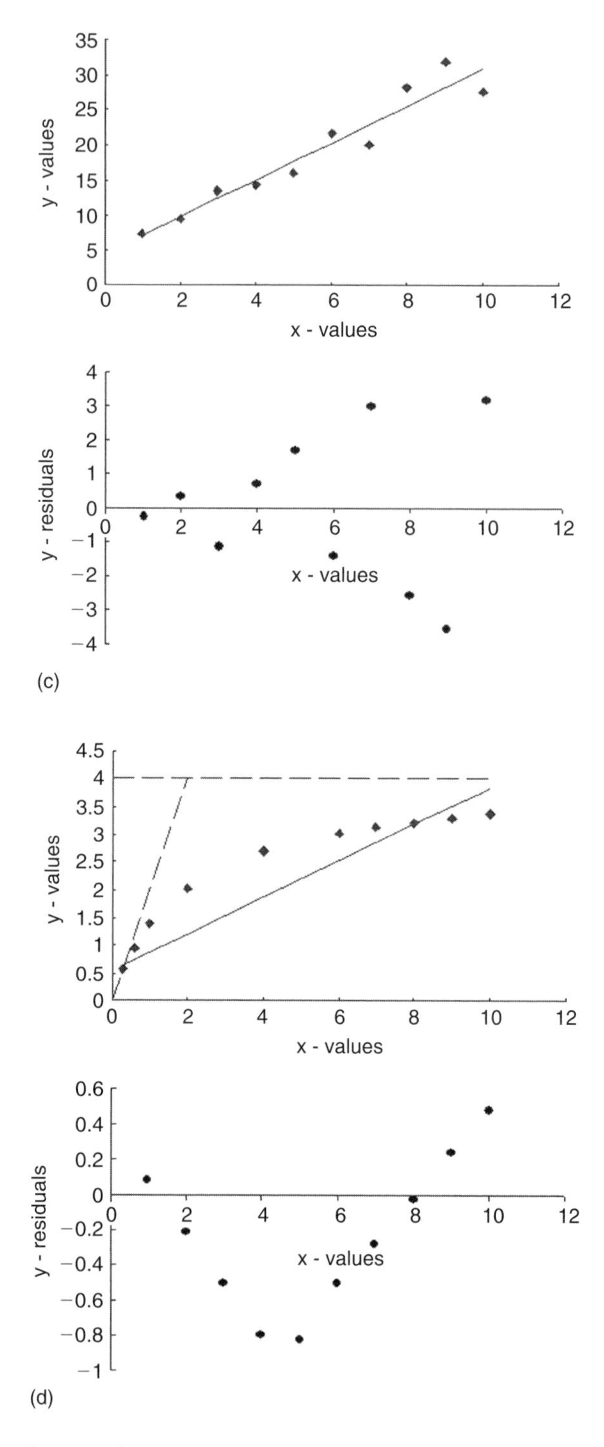

Figure 5.5 Continued.

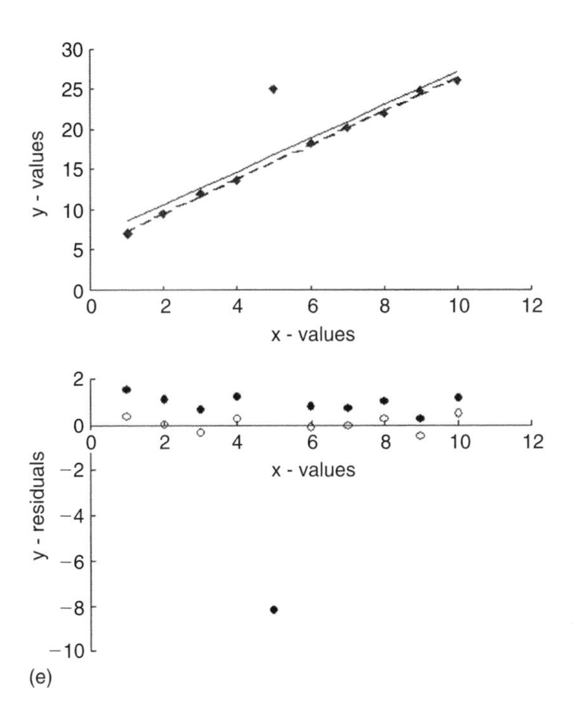

Figure 5.5 Continued.

Table 5.2 Calibration of an absorbance method for the analysis of calcium in milk

$[Ca^{2+}]$ (ppm)	4.00	10.0	15.0	20.0	25.0	30.0	35.0	40.0
Absorbance	0.127	0.281	0.400	0.515	0.580	0.629	0.730	0.779

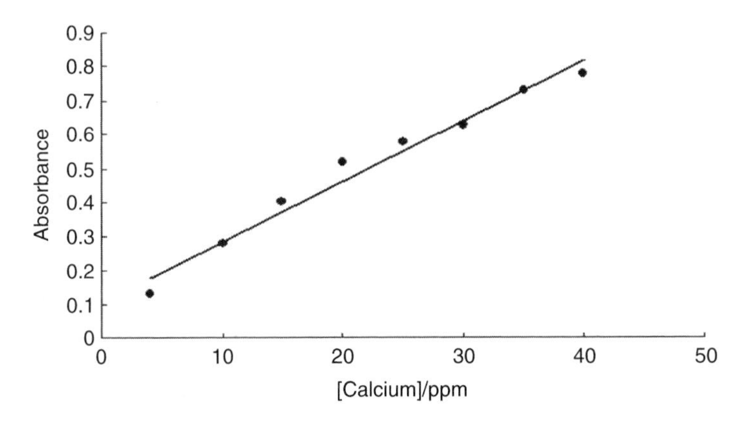

Figure 5.6 Calibration of the analysis of calcium by a spectrophotometric method.

Solution

To ascertain the suitability of the linear calibration model, plot the residual versus concentration. Remember, the residual is given by $(y_i - \hat{y}_i)$, where y_i is the measured response of the AAS instrument for a given calibrator concentration x_i and $\hat{y}_i$ is the estimated value from the regression equation. The calibration data can be expressed in an Excel spreadsheet, as shown in spreadsheet 5.2.

The plot of the residual versus concentration is shown in figure 5.7. There is a clear V-shape to the data. At low concentrations the experimental points lie below the estimated straight line

Spreadsheet 5.2

	A	B	C	D
1	C mg/ml	Absorb	y_hat	Residual
2	4	0.127	0.179	0.0517
3	10	0.281	0.2852	0.0039
4	15	0.400	0.3737	-0.0263
5	20	0.515	0.4622	-0.0528
6	25	0.580	0.5507	-0.0293
7	30	0.629	0.6392	0.0102
8	35	0.730	0.7277	-0.0023
9	40	0.779	0.8162	0.0375

=TREND ($B$2:$B$9, $A$2:$A$9,A2,1)

=(C2-B2)

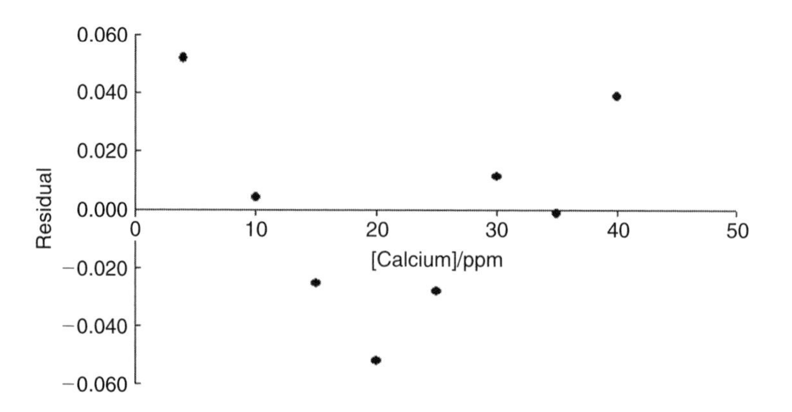

Figure 5.7 Residuals plotted against concentration of calcium for the data in example 5.2.

(positive residuals); they then rise above the line (negative residuals) and then fall back again. This is a clear signature of a curvature in the original plot.

Comment

Looking at the original calibration plot there certainly seems to be well-defined curve in the data rather than a random distribution of calibration points around the regression line and hence a residual plot with a clear trend in the data is not too surprising. Note, however, it is not always easy to tell and hence a residual plot is very useful. Take the calibration plot in example 5.1. A close look at the plot suggests that there, too, may be a gentle curvature in the calibration data. The residual plot in figure 5.8 shows reasonable scatter of data and hence there is no reason to reject the linear calibration model.

5.4 Calibration in Excel

5.4.1 Plotting calibration graphs

The Chart Wizard is the usual starting point for graphs in Excel, and a number of different formats are on offer. For calibration graphs always choose the "XY (Scatter)" plot, which is about halfway down the menu. Do not choose "Line" which spaces out the points equally.

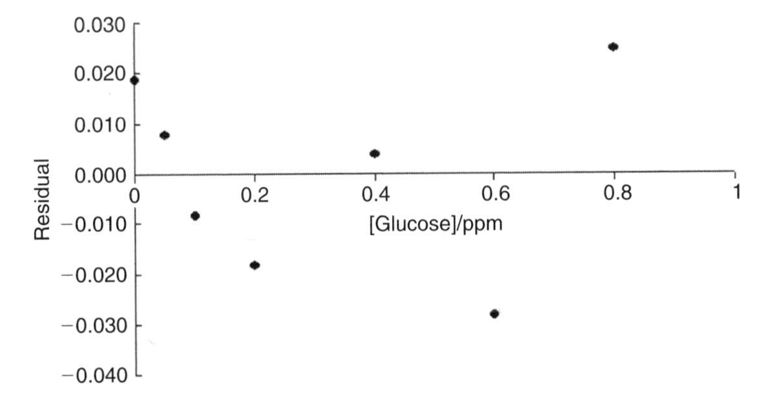

Figure 5.8 Residuals plotted against concentration of glucose for the data in example 5.1.

Following through the dialog boxes of the Wizard leads to a graph with labeled axes, points and/or lines, and a graph title and legend. Further manipulation is possible by clicking on different parts of the graph and by calling up items from the Chart menu, which appears if an existing chart is in focus (i.e., you have clicked on it). Everyone has a preferred way of displaying a chart and we encourage readers to find theirs.

5.4.2 Trendline

Right click on a point in the series and "Add Trendline" will be one of the options. Different fits are on offer including, as default, a simple straight line. By clicking on Options, the equation of the line may be displayed on the graph along with r^2. This is a dangerous option. Do not use Trendline to obtain the equation for the line, as the coefficients are displayed to limited precision. For serious work use the functions described below and if desired only use Trendline to display the fitted line on a calibration graph. Note that the use of Trendline automatically chooses the whole range of points in the series. If you have decided that one point is to be omitted from further consideration then you will have to calculate the fitted line separately using the functions given below and then graph these as a separate series. Remember, Trendline is a quick way of displaying a fitted line on a graph – nothing more.

5.4.3 Functions

There are many functions to help fit data, particularly for linear regression. =SLOPE(*y-range*, *x-range*) and =INTERCEPT(*y-range*, *x-range*) calculate their eponymous coefficients. =TREND(*y-range*, *x-range*, *x*, *const*) calculates the dependent variable (y) at a given independent variable (x) from a linear regression of the ranges specified (see example above). The input *const* is TRUE if there is to be an intercept and FALSE if not (i.e., the graph is forced through zero). As TREND calculates y from x it is not of particular use in calibration and analysis, and only finds employment for creating data for the linear fitted line for graphing, as an alternative to Trendline, or for generating data for residual plots.

Of greatest use, although not so easy to use, is LINEST which in the command line of Excel has the form =LINEST(*y-range*, *x-range*, const, stats). In analysis of calibration data *y-range* is the values of the instrument response, y_i, for a given range of calibrator concentrations, x_i, which is the *x-range*. The constant, *const*, is set to TRUE to have the value of the intercept a calculated. If a is to be zero then FALSE is input for *const*. The input *stats* is another flag which if TRUE creates a number of useful statistics of a linear regression. LINEST is an array function which means its output is over a number of cells. Once you know how to perform a LINEST analysis in Excel, determining the uncertainty in a linear regression certainly requires much less effort than the first principles approach used in example 5.1. This is illustrated below in example 5.3 for the same data used in example 5.1.

Example 5.3

Problem

To determine the standard deviations in the calibration parameters a and b for the calibration of a spectroscopic glucose oxidase enzyme assay, where the calibration data are given in table 5.3.

Solution

The values and uncertainty in the calibration constants will be determined using the array function LINEST. Note these data are the same data as in example 5.1 and therefore the values determined by LINEST can be compared to the values obtained from first principles in example 5.1.

1. Set up a spreadsheet with the values of x_i and y_i in columns.
2. Left click the mouse on a cell and drag across another cell and down five cells. When you lift your finger from the

Table 5.3 Calibration of the analysis of glucose by an enzyme spectroscopic method

[Glucose] (mM)	0.000	0.050	0.100	0.200	0.400	0.600	0.800
Absorbance	0.000	0.057	0.119	0.221	0.383	0.599	0.730

Spreadsheet 5.3

	A	B	C	D	E
1	[G] /mM	A			
2	0	0			
3	0.05	0.057			
4	0.1	0.119			
5	0.2	0.221			
6	0.4	0.383			
7	0.6	0.599			
8	0.8	0.73			
9					
10					
11					
12					
13					
14					
15					
16					

mouse the block of cells two columns wide by five rows deep will remain highlighted, as shown in spreadsheet 5.3.

3. Type =LINEST(in the command line and now go to the start of the y range of data, left click, and drag down the column. The range will appear in the function bar, for example =LINEST(B2:B8. Let go of the mouse button and type a comma. Now left click and drag down the x range of data, and again finish off with a comma. The entry should now look like =LINEST(B2:B8,A2:A8,. Finish off with 1,1) or 0,1) depending on whether you want an intercept. The number 1 is equivalent to TRUE, and quicker to type. The final function in our example is =LINEST(B2:B8, A2:A8,1,1). To complete the output of the array, hold down Shift and Ctrl and press Enter. The 2×5 block will fill with numbers. If you just press Enter by mistake and a single number appears in the top left hand cell, highlight the block again, click in the formula box on the toolbar, and press Shift-Ctrl-Enter again. The spreadsheet now looks like spreadsheet 5.4. If you look carefully at these numbers and compare them to the output spreadsheet from example 5.1 you can work out that the numbers in cells B10 and C10 are the calibration coefficients b and a, respectively. The numbers in cells B11 and C11 are the standard deviations of b and a, respectively, that is, s_b and s_a. The number in cell

Spreadsheet 5.4

	A	B	C	D	E
1	[G] /mM	A			
2	0	0			
3	0.05	0.057			
4	0.1	0.119			
5	0.2	0.221			
6	0.4	0.383			
7	0.6	0.599			
8	0.8	0.73			
9					
10		0.9196636	0.0188176		
11		0.0284681	0.0118481		
12		0.9952318	0.0211536		
13		1043.6201	5		
14		0.4669921	0.0022374		
15					
16					

B12 is the r^2 value (see section 5.5 for a discussion as to why you should not use the r^2 value as a parameter demonstrating the goodness of fit of your linear regression), in cell C12 is the value of $s_{y/x}$, and in cell C13 the degrees of freedom. These values and the last row of the array will be discussed further in the comments section of this example.

Answer

The values of the coefficients of the calibration equation and their associated standard deviations are $a = 0.02$, $b = 0.92\,\text{mM}^{-1}$, $s_a = 0.01$, $s_b = 0.03\,\text{mM}^{-1}$.

Comments

1. Hence the output from LINEST gives the same values as the first principle analysis in example 5.1, but it takes far less time to implement.
2. Note that an alternative to inputting the =LINEST function in the command line is to select your 2 × 5 array of cells for the output, then to use the function option (from the Insert menu, or toolbar icon f_x) and select LINEST. A dialogue box similar to that shown in spreadsheet 5.5 appears and you can enter the ranges for y and x and the constant and *stats* in the appropriate spaces. You still need to press Ctrl-Shift-Enter to see all the output in the 5 × 2 array.

Spreadsheet 5.5

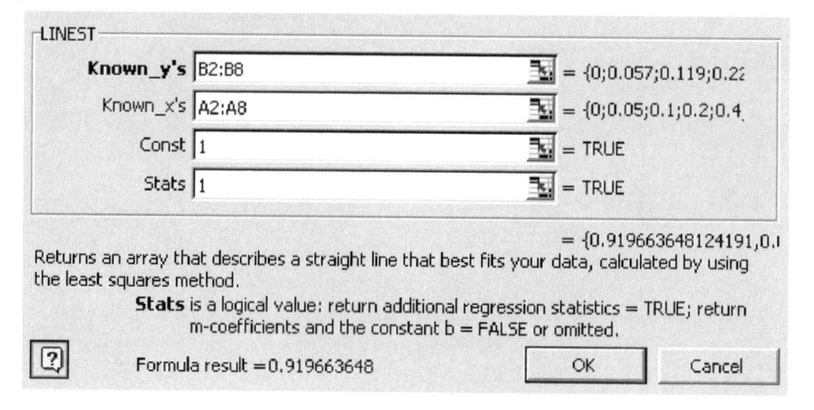

Table 5.4 Output of the Excel LINEST function

Slope: b	Intercept: a
Standard deviation of slope: s_b	Standard deviation of intercept: s_a
Coefficient of determination: r^2	Standard error of regression: $s_{y/x}$
Fisher F-statistic: $F = \frac{\overline{SS}_{\text{regression}}}{\overline{SS}_{\text{residual}}}$	Degrees of freedom of the regression: $df = n-1$ or $n-2$
Sum of squares due to regression: $SS_{\text{regression}}$	Sum of squares due to residual: SS_{residual}

3. The results in the array are the numbers representing the parameters (table 5.4).

The number of degrees of freedom is $n-2$ if the intercept is calculated or is $n-1$ if a is set to zero. The residual sum of squares, SS_{residual}, is

$$SS_{\text{residual}} = \sum_{i=1}^{i=n} (y_i - \hat{y}_i)^2 = df \times s_{y/x}^2 \tag{5.17}$$

SS_{residual} divided by the degrees of freedom is also known as the mean square residual and is equal to the square of the standard error of the regression:

$$\overline{SS}_{\text{residual}} = \frac{SS_{\text{residual}}}{df} = s_{y/x}^2 \tag{5.18}$$

The sum of squares due to the regression is

$$SS_{\text{regression}} = \sum_{i=1}^{i=n} (\bar{y} - \hat{y}_i)^2 \quad (5.19)$$

The degrees of freedom of $SS_{\text{regression}}$ is the number of terms in the calibration containing the dependent variable—here there is just one term (bx), so $SS_{\text{regression}} = \overline{SS}_{\text{regression}}$. If this looks like the ANOVA calculations then do not be surprised to learn we can compare the mean square of the regression with the mean square of the residuals to give an F-statistic. This F compares the mean square arising from the regression to that of the residual. If it is significantly greater than 1, then the regression is significant, which means there is a linear relationship between the dependent and independent variable. As we usually know the fact that we are dealing with a bona fide straight line, the F-statistic is not of great use to the analyst:

$$F = \frac{\overline{SS}_{\text{regression}}}{\overline{SS}_{\text{residual}}} \quad (5.20)$$

This is the value in the LINEST table. The associated probability may be calculated from =FDIST(*F*,*1*,*df*). However, as explained above, it would be a terrible thing if our calibration did not lead to a significant F, as we know that the model does fit the data well (or we would not be using it for calibration!).

If we need only one of the LINEST outputs in a calculation, it may be extracted by the function =INDEX(array, row, column), without having to display the entire array. For example, the standard error of the regression is =INDEX(LINEST(*y-range, x-range, const, stats*),3,2) because $s_{y/x}$ is in the third row, second column of LINEST.

5.5 r^2: A Much Abused Statistic

Calculators and spreadsheets invariably offer a statistic called r or r^2. This is hardly ever of use to an analytical chemist and should not be calculated or quoted to support the quality of a regression. The statistic r is called the correlation coefficient and takes values from -1 to $+1$, and its square, r^2, the coefficient of regression, is the fraction of the variance in the dependent variable explained by the relationship with the independent variable. It has values between 0 (no variance explained by the model) and $+1$ (all variance explained by the model). The correlation coefficient is much used in sciences where relationships between variables subject to many influence factors are being studied. An r^2 of 0.5 might turn out to be highly significant in the case of the epidemiology of some disease. For the analytical chemist, the calibration model is usually a very good description of the relationship between variables. When we construct a linear calibration equation, we do not test that linearity with it, but observe the random scatter of data about the relation. This is built into the assumptions we make in using classic least squares. It would be unusual to see an r^2 of less than 0.9, and frequently 0.999 is considered a minimum. Unfortunately, at this linearity the coefficient of regression is a very poor indicator of the quality of the regression: some quite passable curves have excellent r^2 values. Take the data in example 5.2 as a demonstration of the uselessness of r^2 for calibration data. We can see clearly that a linear calibration model is not valid, and this is confirmed by the residual plot. However, the r^2 value is 0.9756, which certainly seems acceptable until you look at the actual data. The standard error of the regression is the best general statistic and ultimately the uncertainty on a concentration derived from the calibration is of major interest to the analyst.

5.6 The Well-Tempered Calibration

1. When choosing the range of concentrations to construct the calibration relation, it is always good to choose the shortest range possible for the likely range of unknown samples. There is nothing to be gained by calibrating over some enormous range then only using a small part of it. If the unknowns are spread over a wide range it may be better to perform separate calibrations over restricted ranges.
2. Choose sufficient calibration concentrations to cover the range and give a suitable confidence interval on measurement

results. Six should be a minimum number of calibration concentrations with at least duplicate measurements of the unknowns. Evenly space the concentrations throughout the calibration range.

3. Make up calibration solutions independently from reference materials with traceable purities or concentrations. Serial dilutions of a stock solution should be avoided.
4. Choose the solutions for measurement randomly. Do not start at the least concentrated solution and work up to the most concentrated.
5. Inspect a residual plot and recalculate the regression, if necessary, after remeasuring solutions that give apparent outliers and after defining the linear range.
6. Calculate and quote the standard error of the regression. Calculate 95% confidence intervals on any concentration calculated from the calibration relation. If required, quote the regression equation with standard deviations (or 95% confidence intervals) on slope and intercept.
7. Do not forget units! The units of the intercept and its uncertainty are those of y, and for the slope and its uncertainty the units are those of y/x.

5.7 Standard Addition

Standard addition is a method of analysis in which a measurement is made on the sample followed by a second measurement after a known amount of a calibration material is added to the sample.

Suppose the response follows a linear relation with concentration. Let the response of the instrument to the sample solution containing analyte of concentration x_0 before addition be y_0. If a spike is added to bring the added concentration to x_1 with new response y_1, then

$$\begin{aligned} y_0 &= b x_0 \\ y_1 &= b(x_0 + x_1) \end{aligned} \tag{5.21}$$

Rearranging for x_0

$$x_0 = \frac{x_1 y_0}{y_1 - y_0} \tag{5.22}$$

If the spike is added as a solution, it will also dilute the existing analyte and so the equation must reflect this. Let the initial volume of analyte be V and an addition be made of V' of the spike, concentration x_1. The original analyte is diluted by $V/(V+V')$ and the added spike by $V'/(V+V')$. Now the equations are

$$\begin{aligned} y_0 &= bx_0 \\ y_1 &= b\left(x_0 \frac{V}{V+V'} + x_1 \frac{V'}{V+V'}\right) \end{aligned} \tag{5.23}$$

and

$$x_0 = \frac{y_0 x_1 (V'/(V+V'))}{y_1 - y_0(V/(V+V'))} \tag{5.24}$$

Statistics can be applied to repeated determinations of x_0 if sufficient sample is available. More information can be obtained by repeated additions of the standard solution, when the instrumental response is plotted against the concentration of the added standard in the test solution (see figure 5.9).

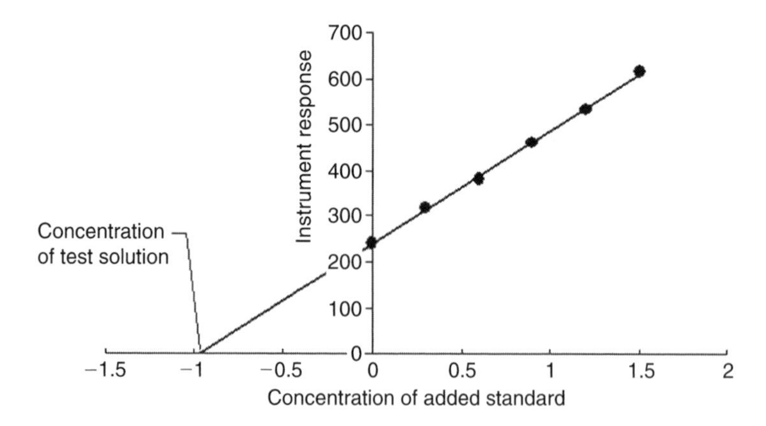

Figure 5.9 Instrument response as a function of added calibration material in analysis by standard addition.

Assuming a linear response with measurand, an estimate of the concentration of analyte in the test solution is

$$\hat{x} = \frac{a}{b} \tag{5.25}$$

where a is the intercept and b the slope and $\hat{x}$ has the same units as the aliquot of the standard solution added. In terms of a graph of instrument response against concentration of added standard the estimate of x happens to be the negative intercept on the concentration axis and may be calculated as the intercept divided by the slope. If there have been n measurements made ($n > 2$), the standard deviation of the concentration estimate may be calculated from the regression line through the points of figure 5.9:

$$s_{\hat{x}} = \frac{s_{x/y}}{b}\sqrt{\frac{1}{n} + \frac{\overline{y}^2}{b^2 \sum_i (x_i - \overline{x})^2}} \tag{5.26}$$

From $s_{\hat{x}}$ the 95% confidence interval on the estimate may be determined by multiplication by the appropriate t-value ($t_{0.05'',n-2}$).

Standard addition is used when there are potential interferents that would lead to a systematic error that is proportional to concentration. Calculation of the concentration by the standard addition method causes these errors in the measurements to cancel. It is also useful if the analyte cannot be extracted from its matrix, and there is not a matrix matched calibrant available. This may be the case in environmental analysis. Note, however, that standard addition does not compensate for a constant additive interferent.

Example 5.4

Problem

Determine the concentration of glucose in a wine sample and its associated uncertainty as a 95% confidence interval by the standard addition method using the enzyme spectroscopic assay of example 5.1. An aliquot of 200 µL of wine is added to a 5 mL volumetric flask to which the reagents for the enzyme assay are added and then the flask is filled to the mark with buffer. The

Table 5.5 Analysis of glucose by standard addition

Added glucose, [glucose] (mM)	0.000	0.050	0.100	0.200	0.300	0.400
Absorbance	0.244	0.284	0.321	0.431	0.550	0.674

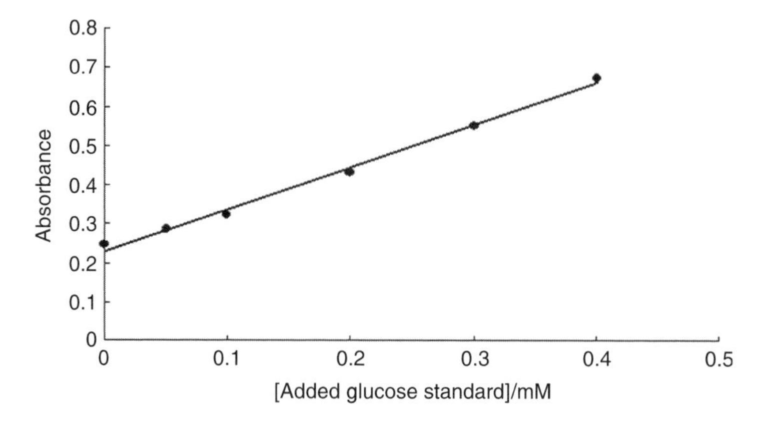

Figure 5.10 Standard addition for glucose concentration in a wine sample in example 5.4.

absorbances of solutions containing different concentrations of added glucose standard are shown in table 5.5.

Solution

1. The first step is to plot the data (figure 5.10) and then perform a regression analysis of the calibration data and the uncertainty in the calibration coefficients a and b. This is best performed using LINEST as shown in spreadsheet 5.6. LINEST indicates the values of a and b are 0.2274 and $1.085\,\text{mM}^{-1}$, respectively, with the standard deviations 0.0095 for s_a and $0.042\,\text{mM}^{-1}$ for s_b.
2. The second step is to estimate the concentration of glucose in the test sample. This is calculated using

$$\hat{x} = \frac{a}{b} = \frac{0.2274}{1.0855} = 0.2095\,\text{mM}$$

3. Determine the uncertainty in the estimate using equation 5.26 where $s_{y/x}$ is obtained from LINEST, and $(x - \overline{x})$ is

Spreadsheet 5.6

	A	B	C
1	[Glucose] /mM	A	x-xbar
2	0	0.244	-0.175
3	0.05	0.284	-0.125
4	0.1	0.321	-0.075
5	0.2	0.431	0.025
6	0.3	0.55	0.125
7	0.4	0.674	0.225
8			
9	0.175	0.417333333	**Means**
10			
11		LINEST	
12		1.085473684	0.22737544
13		0.042296583	0.00949713
14		0.993963263	0.01457545
15		658.6095645	4
16		0.139917558	0.00084978
17			
18	[Glucose] /mM	0.209471166	
19	sx_hat	0.015952752	
20	95% c.l.	0.044292032	

=(A2-$A$9)

=AVERAGE(A2:A7)

=AVERAGE(B2:B7)

=C12/B12

=(C14/B12)*SQRT((1/6)+(B9^2/(B12^2*SUMSQ(C2:C7))))

=TINV(0.05,4)*B19

calculated in spreadsheet 5.6. The spreadsheet shows that $s_{\hat{x}} = 0.0159\,\text{mM}$.

4. Finally calculate the uncertainty as a 95% confidence interval using $t_{0.05'',4}\, s_{\hat{x}} = 2.7765 \times 0.0159 = 0.0443\,\text{mM}$.

Therefore in the 5 mL volumetric flask the concentration of glucose was $0.209 \pm 0.044\,\text{mM}$ (95% confidence interval). Now the 5 mL flask contained diluted wine, so in the original wine sample with a 95% confidence interval [glucose] = $(0.209 \pm 0.044) \times 5/0.2 = 5.24 \pm 0.40\,\text{mM}$.

Answer

The concentration of glucose in the wine as determined using the standard addition method is $5.24 \pm 0.40\,\text{mM}$ (95% confidence interval).

Comments

Note the standard addition method gives a smaller value for the concentration of glucose than in example 5.1 where only a calibration curve was used. The greater value for the calibration curve method suggests there was some interference in the calibration curve which is contributing to the total absorbance measured when analyzing the unknown sample.

There are drawbacks to the use of standard addition. First, the test portion is destroyed by adding a calibrant. Second, the number of analyses for each test material is at least two and is more if it is decided to perform a regression. For a large number of similar samples a separate calibration and the analysis of each test material using that calibration can lead to fewer total measurements. Finally, standard addition does not lead to cancellation of systematic errors if the interferents are in a fixed amount: they just add a constant to the response of the instrument and the result.

5.8 Limits of Detection and Determination

Important characteristics of a method are the limits of detection and determination. The limit of detection is the smallest concentration giving a significant response of the instrument that can be distinguished as being present to above the blank or background response. The limit of determination is the smallest amount of measurand that can be measured with a stated precision.

If a blank material, that is, the matrix of the test material without the analyte, can be analyzed a number of times, the limit of detection is often defined as three times the standard deviation of this blank determination. The limit of detection of the instrumental response is therefore $y_B + 3s_B$, where the subscript B refers to a blank determination. The corresponding concentration is then calculated from the calibration equation (equation 5.4), if it may be assumed that the equation is valid down to that concentration:

$$\hat{x}_{DL} = \frac{y_B + 3s_B - a}{b} \qquad (5.27)$$

It may be not possible to make a measurement in the absence of the analyte. In this case it is reasonable to substitute the intercept of the calibration equation for the blank response (after all, it is supposed to be the response when the concentration is zero) and the standard error of the regression, $s_{y/x}$ for the standard deviation of the blank. In equation 5.27, therefore, $y_B = a$ and $s_B = s_{y/x}$, which gives

$$\hat{x}_{DL} = \frac{3s_{y/x}}{b} \tag{5.28}$$

Equation 5.28 has the advantage that it is calculated entirely from the calibration equation. A more statistically defensible equation from calibration data has been published by ISO (ISO 11843-2:2000, ISO, Geneva):

$$\hat{x}_{DL} = \frac{2t_{0.05',n-2}s_{y/x}}{b}\sqrt{\frac{1}{K} + \frac{1}{I \times J} + \frac{\overline{x}^2}{J\sum_i (x_i - \overline{x})^2}} \tag{5.29}$$

Here, a calibration is performed with I independent calibration materials (including a blank if possible and a calibrator having a value near the expected detection limit) each measured J times. K is the number of replicate measurements that will be done on each test solution to give an average response. (If you are willing to do more repeats you are more likely to pick up the presence of the analyte at small concentrations.) Note that as the t-statistic limits to the value 1.64 for large n, this equation multiplies $s_{y/x}/b$ by at least 3.3 (for $K = 1$), and so gives somewhat greater detection limits than equation 5.28.

Example 5.5

The calibration data of an electrode for the selective detection of copper in water samples, where the electrode was prepared by modifying a gold electrode with cysteine, is shown in table 5.6.

Table 5.6 Calibration of an electrode for the determination of copper near the detection limit

$[Cu^{2+}]$ (nM)	0.0	3.1	6.3	12.6	15.2	20.5	26.8
I ($\mu A\ cm^{-2}$)	0.8	2.8	4.9	8.3	10.2	12.9	16.4

The standard deviation of the blank, s_B, is 0.2 nM ($n = 3$).

Problem

From the calibration data determine the detection limit for copper of the cysteine-modified electrode for a single measurement of a test solution.

Solution

1. Plot the calibration data and determine the calibration parameter and associated uncertainties using LINEST. The calibration plot shown in figure 5.11 confirms the linearity of the data. The results table from LINEST for this data is shown in spreadsheet 5.7
2. The detection limit can be calculated using equation 5.29 where

$$\begin{aligned}\hat{x}_{\mathrm{DL}} &= \frac{2t_{0.05',n-2}s_{y/x}}{b}\sqrt{\frac{1}{K}+\frac{1}{IJ}+\frac{\overline{x}^2}{J\sum_i (x_i-\overline{x})^2}} \\ &= \frac{2\times 2.015\times 0.1441}{0.5802}\sqrt{\frac{1}{1}+\frac{1}{7}+\frac{(12.07)^2}{557.6}} \\ &= 1.865\,\mathrm{nM}\end{aligned}$$

The input values can be obtained from spreadsheet 5.8.

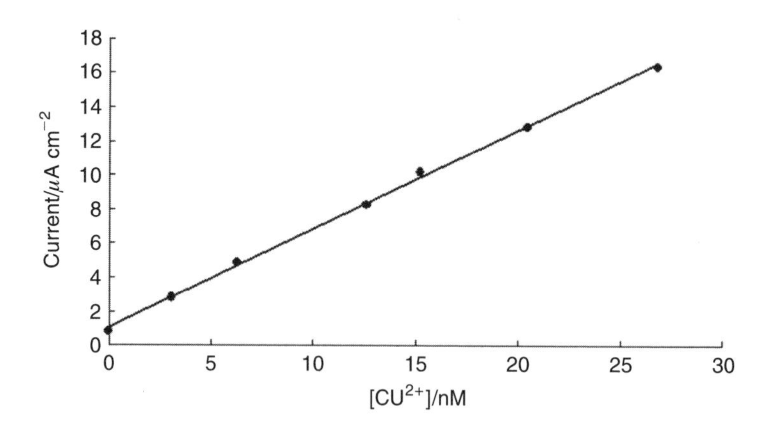

Figure 5.11 Calibration of copper anodic stripping voltammetry experiment in example 5.5.

Spreadsheet 5.7

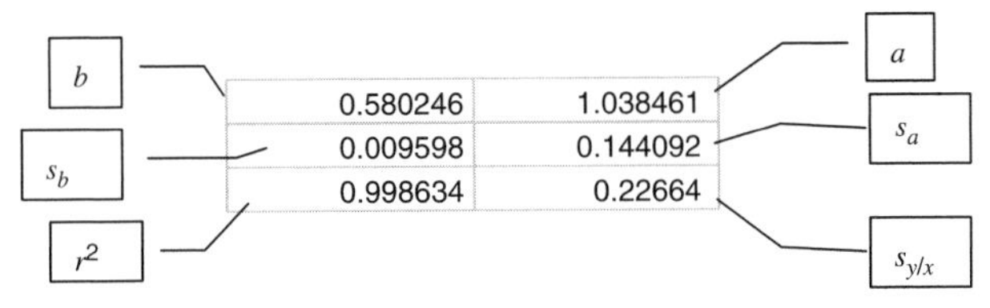

Spreadsheet 5.8

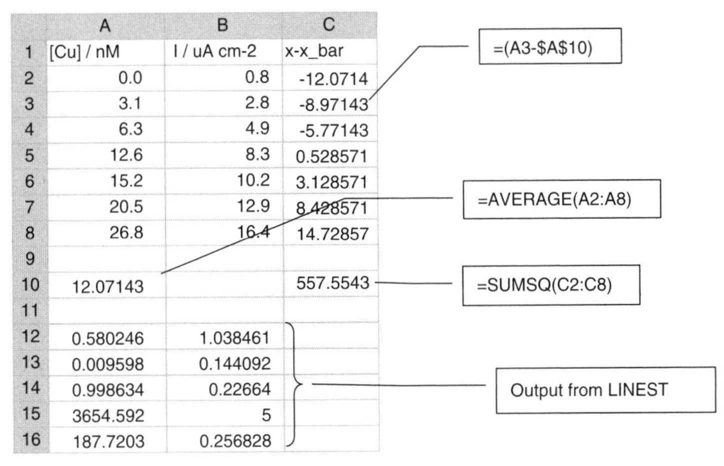

	A	B	C
1	[Cu] / nM	I / uA cm-2	x-x_bar
2	0.0	0.8	-12.0714
3	3.1	2.8	-8.97143
4	6.3	4.9	-5.77143
5	12.6	8.3	0.528571
6	15.2	10.2	3.128571
7	20.5	12.9	8.428571
8	26.8	16.4	14.72857
9			
10	12.07143		557.5543
11			
12	0.580246	1.038461	
13	0.009598	0.144092	
14	0.998634	0.22664	
15	3654.592	5	
16	187.7203	0.256828	

Answer

The detection limit of the cysteine-modified electrode for copper in water samples is 1.87 nM.

Comments

1. A more simple, but less statistically defensible, equation is equation 5.28. $3s_{y/x}/b$ is $3 \times 0.2266/0.5802 = 1.17\,\text{mM}$. This underestimates the detection limit.
2. As we have a number of blank measurements an alternative to using equation 5.29 would be to use equation 5.27:

$$\hat{x}_{\text{DL}} = \frac{y_{\text{B}} + 3s_{\text{B}} - a}{b} = \frac{0.80 + 3 \times 0.20 - 1.04}{0.58} = 0.62\,\text{nM}$$

The lower detection limit obtained using this equation is certainly appealing, as is the simplicity in using this equation, but it is important to emphasize that equation 5.29 is the more statistically defensible and more conservative method of calculating the detection limit. The problem lies in only using three measurements of the blank (thus s_{blank} is not a good estimate of σ_{blank}), and the accident that the blank response happened to be smaller than the intercept.

3. If you need to establish the detection limit then this should be done with some care. A calibration with blank and a solution with near the expected detection limit should be done, with the calculation of equation 5.29 ensuring that the chance of both errors (deciding that there is a detectable concentration when there is not, and missing the presence of a detectable concentration) is about 5%.

If the detection limit is really important for a particular application it is always a good idea to analyze a test solution containing the estimated detection limit concentration. This is the only way that the capabilities of the method can be shown for sure.

The limit of determination (as distinct from the limit of detection) is even more of a movable feast, as it depends on the required precision. A proposal that this limit be calculated as $y_B + 10s_B$ has not found great favor. A suitable level really depends on the requirements of the analysis. For some measurements a rather poor precision may be "fit for purpose," while in others extreme precision may be necessary. The concept of "target value for uncertainty" (TVU) or "target measurement uncertainty" (TMU) has recently been adopted in a number of fields. Here, the client, or where a TMU is specified for a method to be used regularly for a particular purpose an independent authority, specifies what the largest acceptable measurement uncertainty will be for a given set of measurements. For example, a maximum relative standard deviation of 0.5% might be set for measurements of radioactive waste. This is policed using interlaboratory proficiency tests, in which a sample of known concentration is sent to each of a number of participating laboratories, and each laboratory is required to achieve the set measurement uncertainty as a demonstration of its capability.

Appendix

The critical values of different statistics presented here have been generated in Microsoft Excel, using inbuilt functions and other formulae.

Table A.1 Two-tailed Student *t*-values (=TINV(α, *df*))

	Confidence Interval			
	90%	*95%*	*99%*	*99.9%*
Degrees of Freedom	$\alpha = 0.10$	$\alpha = 0.05$	$\alpha = 0.01$	$\alpha = 0.001$
1	6.31	12.7	63.7	637
2	2.92	4.30	9.92	31.6
3	2.35	3.18	5.84	12.9
4	2.13	2.78	4.60	8.61
5	2.02	2.57	4.03	6.87
6	1.94	2.45	3.71	5.96
7	1.89	2.36	3.50	5.41
8	1.86	2.31	3.36	5.04
9	1.83	2.26	3.25	4.78
10	1.81	2.23	3.17	4.59
11	1.80	2.20	3.11	4.44
12	1.78	2.18	3.05	4.32
14	1.76	2.14	2.98	4.14
16	1.75	2.12	2.92	4.01
18	1.73	2.10	2.88	3.92
20	1.72	2.09	2.85	3.85
30	1.70	2.04	2.75	3.65
50	1.68	2.01	2.68	3.50
∞	1.64	1.96	2.58	3.29

Table A.2 One-tailed Student t-values. As TINV only gives two-tailed values, we must multiply α by 2 to calculate the correct value, i.e., =TINV($2 \times \alpha$, df)

	Confidence Interval			
Degrees of Freedom	*90%* $\alpha = 0.10$	*95%* $\alpha = 0.05$	*99%* $\alpha = 0.01$	*99.9%* $\alpha = 0.001$
1	3.08	6.31	31.82	318.29
2	1.89	2.92	6.96	22.33
3	1.64	2.35	4.54	10.21
4	1.53	2.13	3.75	7.17
5	1.48	2.02	3.36	5.89
6	1.44	1.94	3.14	5.21
7	1.41	1.89	3.00	4.79
8	1.40	1.86	2.90	4.50
9	1.38	1.83	2.82	4.30
10	1.37	1.81	2.76	4.14
11	1.36	1.80	2.72	4.02
12	1.36	1.78	2.68	3.93
14	1.35	1.76	2.62	3.79
16	1.34	1.75	2.58	3.69
18	1.33	1.73	2.55	3.61
20	1.33	1.72	2.53	3.55
30	1.31	1.70	2.46	3.39
50	1.30	1.68	2.40	3.26
∞	1.28	1.64	2.33	3.09

Table A.3 Values of $G_{critical}$ used for Grubbs's test for outliers, calculated as =(n−1)/SQRT(n)*SQRT((TINV(α/n, n−2))^2/(n−2+TINV(α/n, n−2)^2))

	Confidence Level			
Number of Data, n	*90%* $\alpha = 0.1$	*95%* $\alpha = 0.05$	*99%* $\alpha = 0.01$	*99.9%* $\alpha = 0.001$
3	1.15	1.15	1.15	1.15
4	1.46	1.48	1.50	1.50
5	1.67	1.72	1.76	1.78
6	1.82	1.89	1.97	2.02
7	1.94	2.02	2.14	2.22
8	2.03	2.13	2.27	2.38
9	2.11	2.22	2.39	2.52
10	2.18	2.29	2.48	2.64
11	2.23	2.35	2.56	2.75
12	2.28	2.41	2.64	2.84
14	2.37	2.51	2.76	3.00
16	2.44	2.59	2.85	3.12
18	2.50	2.65	2.93	3.23
20	2.56	2.71	3.00	3.31
30	2.75	2.91	3.24	3.61
40	2.87	3.04	3.38	3.79
50	2.96	3.13	3.48	3.91

Table A.4 Two tailed Fisher F-values for $\alpha = 0.05$. As Excel calculates one-tailed values, the function used is =TINV(0.025, df_1, df_2)

Degrees of Freedom of Denominator	*Degrees of Freedom of Numerator*																		
	1	2	3	4	5	6	7	8	9	10	11	12	14	16	18	20	30	50	∞
1	161.5	199.5	215.7	224.6	230.2	234.0	236.8	238.9	240.5	241.9	243.0	243.9	245.4	246.5	247.3	248.0	250.1	251.8	254.3
2	18.51	19.00	19.16	19.25	19.30	19.33	19.35	19.37	19.38	19.40	19.40	19.41	19.42	19.43	19.44	19.45	19.46	19.48	19.50
3	10.13	9.55	9.28	9.12	9.01	8.94	8.89	8.85	8.81	8.79	8.76	8.74	8.71	8.69	8.67	8.66	8.62	8.58	8.53
4	7.71	6.94	6.59	6.39	6.26	6.16	6.09	6.04	6.00	5.96	5.94	5.91	5.87	5.84	5.82	5.80	5.75	5.70	5.63
5	6.61	5.79	5.41	5.19	5.05	4.95	4.88	4.82	4.77	4.74	4.70	4.68	4.64	4.60	4.58	4.56	4.50	4.44	4.37
6	5.99	5.14	4.76	4.53	4.39	4.28	4.21	4.15	4.10	4.06	4.03	4.00	3.96	3.92	3.90	3.87	3.81	3.75	3.67
7	5.59	4.74	4.35	4.12	3.97	3.87	3.79	3.73	3.68	3.64	3.60	3.57	3.53	3.49	3.47	3.44	3.38	3.32	3.23
8	5.32	4.46	4.07	3.84	3.69	3.58	3.50	3.44	3.39	3.35	3.31	3.28	3.24	3.20	3.17	3.15	3.08	3.02	2.93
9	5.12	4.26	3.86	3.63	3.48	3.37	3.29	3.23	3.18	3.14	3.10	3.07	3.03	2.99	2.96	2.94	2.86	2.80	2.71
10	4.96	4.10	3.71	3.48	3.33	3.22	3.14	3.07	3.02	2.98	2.94	2.91	2.86	2.83	2.80	2.77	2.70	2.64	2.54
11	4.84	3.98	3.59	3.36	3.20	3.09	3.01	2.95	2.90	2.85	2.82	2.79	2.74	2.70	2.67	2.65	2.57	2.51	2.40
12	4.75	3.89	3.49	3.26	3.11	3.00	2.91	2.85	2.80	2.75	2.72	2.69	2.64	2.60	2.57	2.54	2.47	2.40	2.30
14	4.60	3.74	3.34	3.11	2.96	2.85	2.76	2.70	2.65	2.60	2.57	2.53	2.48	2.44	2.41	2.39	2.31	2.24	2.13
16	4.49	3.63	3.24	3.01	2.85	2.74	2.66	2.59	2.54	2.49	2.46	2.42	2.37	2.33	2.30	2.28	2.19	2.12	2.01
18	4.41	3.55	3.16	2.93	2.77	2.66	2.58	2.51	2.46	2.41	2.37	2.34	2.29	2.25	2.22	2.19	2.11	2.04	1.92
20	4.35	3.49	3.10	2.87	2.71	2.60	2.51	2.45	2.39	2.35	2.31	2.28	2.22	2.18	2.15	2.12	2.04	1.97	1.84
30	4.17	3.32	2.92	2.69	2.53	2.42	2.33	2.27	2.21	2.16	2.13	2.09	2.04	1.99	1.96	1.93	1.84	1.76	1.62
50	4.03	3.18	2.79	2.56	2.40	2.29	2.20	2.13	2.07	2.03	1.99	1.95	1.89	1.85	1.81	1.78	1.69	1.60	1.44
∞	3.84	3.00	2.60	2.37	2.21	2.10	2.01	1.94	1.88	1.83	1.79	1.75	1.69	1.64	1.60	1.57	1.46	1.35	1.03

Table A.5 One-tailed Fisher F-values for $\alpha = 0.05$. Calculated in Excel by =TINV(0.05, df_1, df_2)

Degrees of Freedom of Denominator	Degrees of Freedom of Numerator																		
	1	2	3	4	5	6	7	8	9	10	11	12	14	16	18	20	30	50	∞
1	161.5	199.5	215.7	224.6	230.2	234.0	236.8	238.9	240.5	241.9	243.0	243.9	245.4	246.5	247.3	248.0	250.1	251.8	254.3
2	18.51	19.00	19.16	19.25	19.30	19.33	19.35	19.37	19.38	19.40	19.40	19.41	19.42	19.43	19.44	19.45	19.46	19.48	19.50
3	10.13	9.55	9.28	9.12	9.01	8.94	8.89	8.85	8.81	8.79	8.76	8.74	8.71	8.69	8.67	8.66	8.62	8.58	8.53
4	7.71	6.94	6.59	6.39	6.26	6.16	6.09	6.04	6.00	5.96	5.94	5.91	5.87	5.84	5.82	5.80	5.75	5.70	5.63
5	6.61	5.79	5.41	5.19	5.05	4.95	4.88	4.82	4.77	4.74	4.70	4.68	4.64	4.60	4.58	4.56	4.50	4.44	4.37
6	5.99	5.14	4.76	4.53	4.39	4.28	4.21	4.15	4.10	4.06	4.03	4.00	3.96	3.92	3.90	3.87	3.81	3.75	3.67
7	5.59	4.74	4.35	4.12	3.97	3.87	3.79	3.73	3.68	3.64	3.60	3.57	3.53	3.49	3.47	3.44	3.38	3.32	3.23
8	5.32	4.46	4.07	3.84	3.69	3.58	3.50	3.44	3.39	3.35	3.31	3.28	3.24	3.20	3.17	3.15	3.08	3.02	2.93
9	5.12	4.26	3.86	3.63	3.48	3.37	3.29	3.23	3.18	3.14	3.10	3.07	3.03	2.99	2.96	2.94	2.86	2.80	2.71
10	4.96	4.10	3.71	3.48	3.33	3.22	3.14	3.07	3.02	2.98	2.94	2.91	2.86	2.83	2.80	2.77	2.70	2.64	2.54
11	4.84	3.98	3.59	3.36	3.20	3.09	3.01	2.95	2.90	2.85	2.82	2.79	2.74	2.70	2.67	2.65	2.57	2.51	2.40
12	4.75	3.89	3.49	3.26	3.11	3.00	2.91	2.85	2.80	2.75	2.72	2.69	2.64	2.60	2.57	2.54	2.47	2.40	2.30
14	4.60	3.74	3.34	3.11	2.96	2.85	2.76	2.70	2.65	2.60	2.57	2.53	2.48	2.44	2.41	2.39	2.31	2.24	2.13
16	4.49	3.63	3.24	3.01	2.85	2.74	2.66	2.59	2.54	2.49	2.46	2.42	2.37	2.33	2.30	2.28	2.19	2.12	2.01
18	4.41	3.55	3.16	2.93	2.77	2.66	2.58	2.51	2.46	2.41	2.37	2.34	2.29	2.25	2.22	2.19	2.11	2.04	1.92
20	4.35	3.49	3.10	2.87	2.71	2.60	2.51	2.45	2.39	2.35	2.31	2.28	2.22	2.18	2.15	2.12	2.04	1.97	1.84
30	4.17	3.32	2.92	2.69	2.53	2.42	2.33	2.27	2.21	2.16	2.13	2.09	2.04	1.99	1.96	1.93	1.84	1.76	1.62
50	4.03	3.18	2.79	2.56	2.40	2.29	2.20	2.13	2.07	2.03	1.99	1.95	1.89	1.85	1.81	1.78	1.69	1.60	1.44
∞	3.84	3.00	2.60	2.37	2.21	2.10	2.01	1.94	1.88	1.83	1.79	1.75	1.69	1.64	1.60	1.57	1.46	1.35	1.03

Bibliography

. . . .

There is a wide and extensive literature of applied statistics. There are texts for statistics in every kind of science and engineering, and we have read many of them. The short list below represents books that we feel will add value to what you have learned here. Rather than give an exhaustive list, we have deliberately excluded books that, in our opinion, will not help. Indeed some texts could undo what little good we may have achieved.

Historical

There are some texts that are of historical interest that are still readable today. While not recommending them as "must reads," they often contain nuggets that have been passed over in the retelling by other texts (no doubt including ours). The book by Youden has been reprinted by the National Institute of Standards and Technology (NIST), for which it can be thanked, and can be downloaded for free from http://physics.nist.gov/Divisions/Div844/facilities/phdet/pdf/expmeas.pdf

1. Youden, W. J. (1961). *Experimentation and Measurement*, National Institute of Standards and Technology, Gaithersburg, Md.
2. Box, G., Hunter, W. et al. (1978). *Statistics for Experimenters, An Introduction to Design, Data Analysis and Model Building*, John Wiley, New York.

3. Coombe, C. (1964). *A Theory of Data*, John Wiley, Chichester, UK.

General Statistical Texts

We understand that chemistry is not statistics, and that books about data analysis for chemists might miss the larger statistical points. Here are a couple of books that might describe this greater picture in a way that chemists might understand. The text by Wild is more dense but does relate statistics to the underlying probability theory.

4. Ramsey, F. L. and Schafer, D. W. (2002). *The Statistical Sleuth*, Duxbury Press, Pacific Grove, Calif.
5. Wild, C. J. (2000). *Chance Encounters. A First Course in Data Analysis and Inference*, John Wiley, New York.

Statistics for Chemistry

Until we wrote this text, the book by Miller and Miller was the data analysis book the we recommended in our courses. For a slim volume our students thought it overpriced, and chemometrics introduced in the recent revisions was, in our view, unnecessary, but the early chapters do cover what an analytical chemist needs to know. Apart from chapters in larger analytical textbooks, there is no other useful book on the market.

6. Miller, J. N. and Miller, J. C. (2000). *Statistics and Chemometrics in Analytical Chemistry*, 4th edition, Prentice Hall, Harlow, UK.
7. Meier, P. C. and Zund, R. E. (1993). *Statistical Methods in Analytical Chemistry*, Wiley Interscience, New York.

Data Analysis with Excel

There has been a realization that much of the basic data manipulation may be done in a spreadsheet, and for the present moment in the 21st century this means Microsoft Excel. The Data Analysis ToolPak provides many useful routines that perform the functions described in this book. While we have leaned heavily on the use of spreadsheets we have tried to not let them take over. An alternative approach is to focus on the practical aspects of spreadsheets and teach data analysis from this standpoint. The first book by de Levie listed below is the most comprehensive book on Excel and contains some very useful macros for analytical chemistry. Our friend Les Kirkup, a physicist, has hedged his bets with Excel in the title but the book is a more traditional approach to general scientific data analysis. The text by Billo is now somewhat out of date, although it does cover a wider range of chemical applications.

8. de Levie, R. (2001). *How to Use Excel in Analytical Chemistry and in General Scientific Data Analysis*, Cambridge University Press, Cambridge, UK.
9. de Levie, R. (2004). *Advanced Excel for Scientific Data Analysis*, Oxford University Press, New York.
10. Kirkup, L. (2002). *Data Analysis with Excel®. An Introduction for Physical Scientists*, Cambridge University Press, Cambridge, UK.
11. Billo, E. J. (1997). *Excel for Chemists*, Wiley-VCH, New York.

Chemometrics

Having mastered basic data analysis the world of chemometrics is open to you. Data comes in many shapes and sizes and modern instrumentation gives ever more potential information. Chemometrics provides the tools to unlock that information through a range of mathematical and computational methods. The current "bible" of chemometrics is the two-volume work by Massart et al., which covers all of the material in our text plus much more. It is very direct and, although having good examples, requires careful reading to understand the principles. Despite the many and varied specialist chemometrics books, we mention only one other, a recent book by Brereton, which combines good explanation with a rigorous treatment.

12. Massart, D. L., Vandeginste, B. G. M., Buydens, J. M. C., de Jong, S., Lewi, P. J., and Smeyers-Verberke, J. (1997).

Handbook of Chemometrics and Qualimetrics, Elsevier, Amsterdam.

13. Brereton, R. G. (2003). *Chemometrics: Data Analysis for the Laboratory and Chemical Plant*, John Wiley, Chichester, UK.

Quality Control

There is a nice book from the Royal Society of Chemistry that is directed at the statistics associated with quality assurance in chemical laboratories.

14. Mullins, E. (2003). *Statistics for the Quality Control Chemistry Laboratory*, Royal Society of Chemistry, Cambridge, UK.

Index